O Level Physics: Questions & Answers

Silvanus Oliver

ISBN-10: 1517788048
ISBN-13: 978-1517788049

CONTENTS

Chapter one: Cathode and X-Rays

1. For a given source of X-rays, how would the following be controlled: -
   a. The intensity.

   b. The penetrating power.

   c. The exposure to patients.

Solution:

a. Intensity (or quantity) is controlled by varying the heating current.

b. The penetrating power (quality) is controlled by the accelerating voltage.

c. Exposure to patients – limiting periods of exposure and using lead shields against stray radiations.

2. An accelerating potential of 20KV is applied to an X-ray tube.
   a. Determine the velocity with which electrons strike the target.

   b. State the energy changes that take lace at the target.

Solution:

a. $V = 20KV = 20 \times 10^3 V$

k.e. of electron $= eV = \frac{1}{2}mv^2$

$= 1.6 \times 10^{-19} C \times 20 \times 10^3 = \frac{1}{2}mv^2$

$$V^2 = \frac{2 \times 1.6 \times 10\text{-}19 \times 20 \times 103}{9.1 \times 10^{-31}}$$

$= 8.386 \times 10^7$ m/s

b. *The kinetic energy (mechanical) of bombarding electrons is converted to heat energy, internal energy and most becomes X-rays energy.*

3. The diagram shows an x-ray tube drawn by student. Use it to answer the following questions.

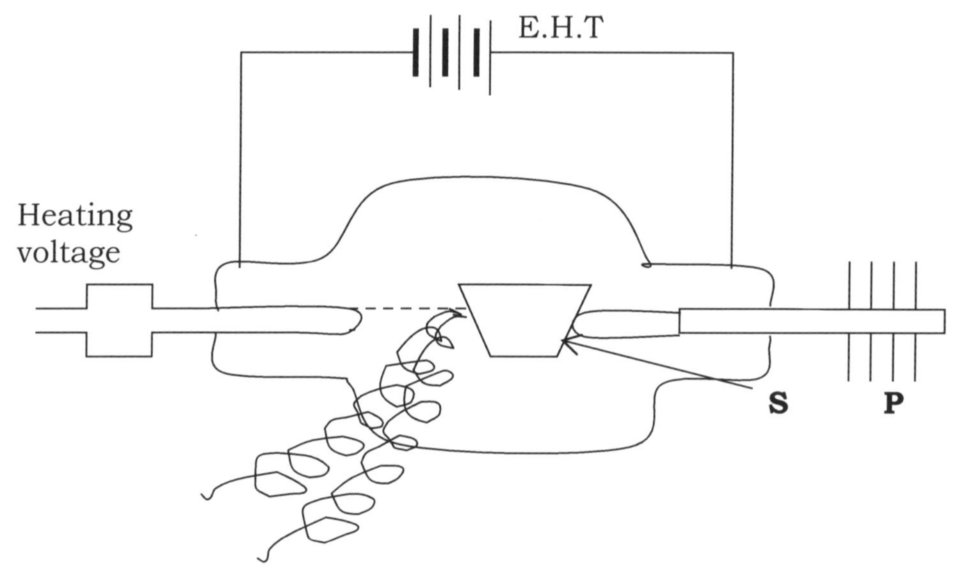

a. State the property possessed by the metal-labelled S.

b. State the function of part labelled P.

Solution:

a. *Has a high melting point.*

b. *Conducts heat away/cools the tube.*

4. State one use of X- rays in each of the following:
   a. Medicine

   b. Industry

Solution:

a. - *Detect internal injuries.*
- *Destroy cancerous cells.*

b. - *Detect flaws in metal castings*
- *Sterilise surgical equipment.*

5. X-rays are emitted when the tube operates at 300V and a current of 0.01A is passing through it. Calculate:
   a. The velocity of the electrons on hitting the target.

   b. Assuming that 10% of energy is converted to X-rays per second, determine the minimum wavelength of the X-rays emitted.

Solution:

a.

$$\tfrac{1}{2}mv^2 = eV$$

$$\Leftrightarrow \quad v = \sqrt{\frac{2eV}{m}}$$

$$= \sqrt{\frac{2 \times 16 \times 10^{-19} \times 300}{9.1 \times 10^{-31}}}$$

$$= 32 \times 10^{7} m/s$$

b.

$$\frac{10}{100} \times eV = \frac{hc}{\lambda}$$

$$\Leftrightarrow \quad \lambda = \frac{hc}{eV} \times \frac{100}{10}$$

$$= \frac{6.63 \times 10^{-34} \times 3 \times 10^{8} \times 100}{1.6 \times 10^{-19} \times 300 \times 10}$$

$$= 4.0 \times 10^{-8} m$$

6. The diagram shows a cathode ray oscilloscope (C.R.O)

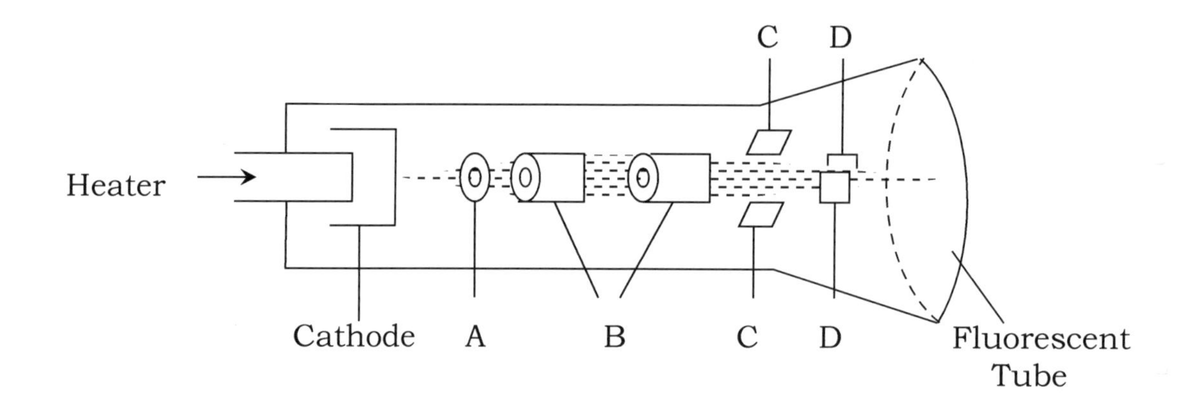

a. Name the parts labelled A and B.

b. What are the functions of the parts labelled C and D?

c. Explain how the electrons are produced.

d. Explain why the tube is evacuated.

Solution:

a. A - Grid
B - Anode
b. C - Y-plate
D - X-plate
c. By thermionic emission – the cathode is heated by the filament of the heater hence electrons break loose.
d. To avoid the electrons colliding with gas molecules and lose their energy.

7. Explain why a television set needs two time base circuits.

Solution:

- *This enables the electron beam to cover the whole screen in a systematic way.*

8. One disadvantage of x-rays production is that 99% of the kinetic energy of thermionic electrons is converted to heat which may melt the target metal. State two methods usually used to prevent this melting.

Solution:

- *The target anode is made of a metal (tungsten) with a high melting point :– 3,380^{0}C*
- *The target area is increased about 100 times by rotating the anode.*

9. State one property of X-rays that is similar to that of visible light.

Solution:

- *They travel at the same velocity.*
- *They are transverse progressive waves*
- *They obey the laws of reflection and refraction.*

10. The figure shows the features of an X-ray tube.

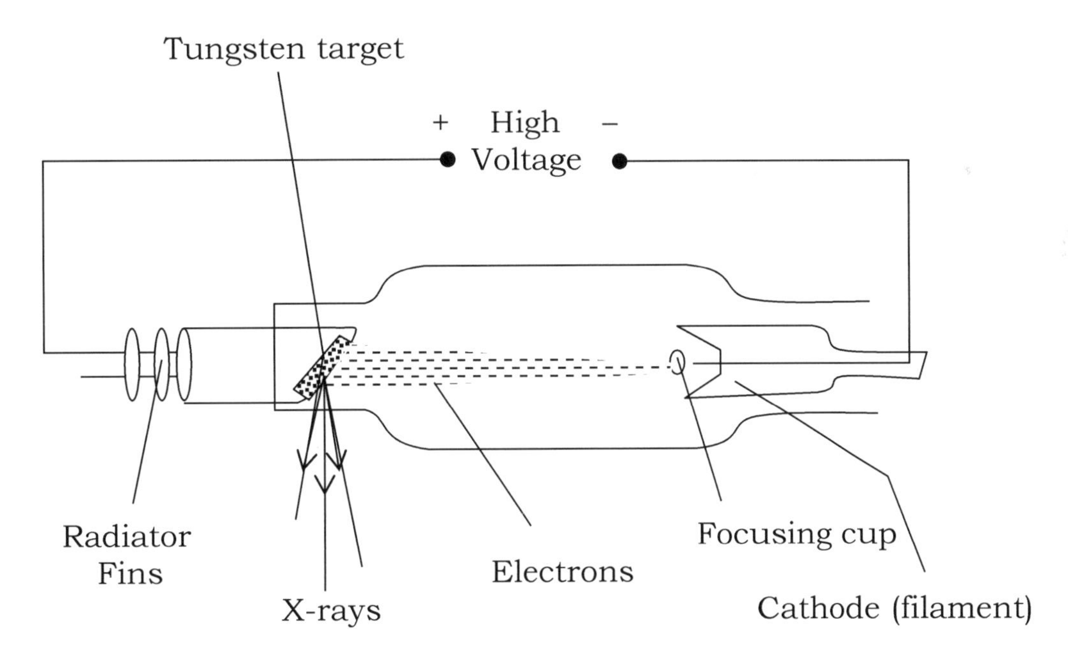

a. Explain how the electrons produced produce X-rays.

b. What is the purpose of the radiator fins connected to the anode?

c. Why is tungsten suitable as the target?

d. Differentiate between hard X-rays and soft X-rays.

e. Give two suitable applications of X-rays.

Solution:

a. The high-energy electrons are made to strike a target, which emits the X-rays as a form of energy.

b. Radiator fins cool the tube by conducting away the heat produced at the anode.

c. Tungsten has a high melting point.

d. Hard X-rays

- *Have high penetrating power*
- *Produced by high velocity electrons*
- *Have shorter wave length*
- *Produced by high accelerating potential.*

Soft X-rays

- *Have low penetrating power*
- *Produced by low velocity electrons*
- *Have a longer wave length*
- *Produced by low accelerating potential.*

e. Uses of X-rays

- *Detect internal injuries of a body*
- *Treatment of malignant growth.*
- *Locate internal cracks of metal bars*
- *Used in crystallography.*

11. A target was bombarded by electrons accelerated by a voltage of 10^6V. If all the kinetic energy of the electrons was converted to X-rays, calculate:

a. The k.e. of the electrons

b. The frequency of the photons emitted

(Charge of an electron = 1.6×10^{-19}C; plank's constant = 6.63×10^{-34}Js)

Solution:

a. $k.e. = \frac{1}{2}mv^2 = eV$

$= 1.6 \times 10^{-19} \times 10^6$

$= 1.6 \times 10^{-13} J$

b. $E = hf$

$\Rightarrow \quad 1.6 \times 10^{-13} = 6.63 \times 10^{-34} \times f$

$\Leftrightarrow \quad f = \dfrac{1.6 \times 10^{-13}}{6.63 \times 10^{-34}} = 2.41 \times 10^{20} Hz$

Chapter two: Current and Mains Electricity

1. a. State the Ohm's Law.

   b. The figure below shows an arrangement of electrical component in a circuit.

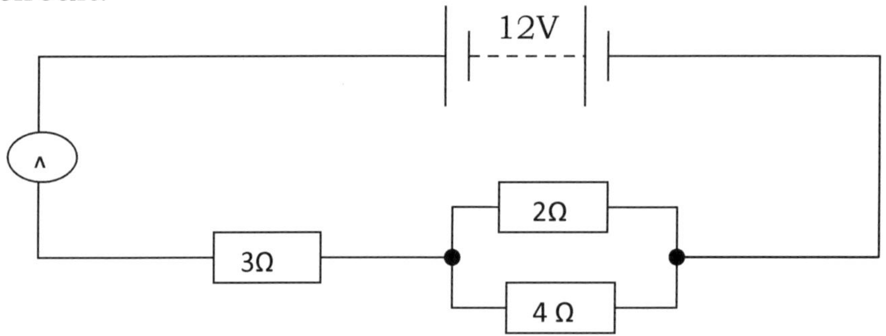

Find:

i. The current through the ammeter.

ii. The current through the 2Ω resistor.

iii. The p.d. across the 4Ω resistor.

Solutions:

a. *The current passing through a conductor at constant temperature is directly proportional to the p.d. between its ends, provided physical conditions remain constant.*

b. *i.* Total resistance $= 3 + \frac{2 \times 4}{2 + 4}$

$= 3 + 1.333 = 4.333\Omega$

$V = IR \Leftrightarrow I = \frac{V}{R} = \frac{12}{4.333} = 2.769A$

ii. p.d. across 3Ω resistor $= 2.769 \times 3$

$= 8.307V$

p.d across parallel resistors $= 12 - 8.307$

$= 3.693V$

Current thro'2Ω resistor $I = \frac{V}{A} = \frac{3.693}{2} = 1.8465A$

iii. p.d. across 4Ω resistor = p.d. across the 2Ω resistor

$= 3.693V$

2. A student was provided with connecting wires, a dry cell, a variable resistor, a switch, an ammeter and a voltmeter. He was required to measure voltage across the cell for various values of current drawn from it.

a. i. Draw a possible circuit diagram for this experiment .

ii. Describe how the results are obtained.

b. The table shows the results obtained in an experiment similar to the one described above

Current I (A)	0.10	0.20	0.30	0.40	0.60	0.80
Voltage V (V)	1.43	1.30	1.19	1.07	0.82	0.58

i. Plot a graph of V (y-axis) against I

ii. Find V^I, the value of V when I = 0.

iii. State what V^I represents.

iv. Determine r, the internal resistance of the cell, given that

$$V = V^I - Ir$$

3. a. State two factors that determine the resistance of a conductor.

b. In the figure below, $R_1 = R_4 = 100\Omega$ and $R_2 = R_3 = 50\Omega$.

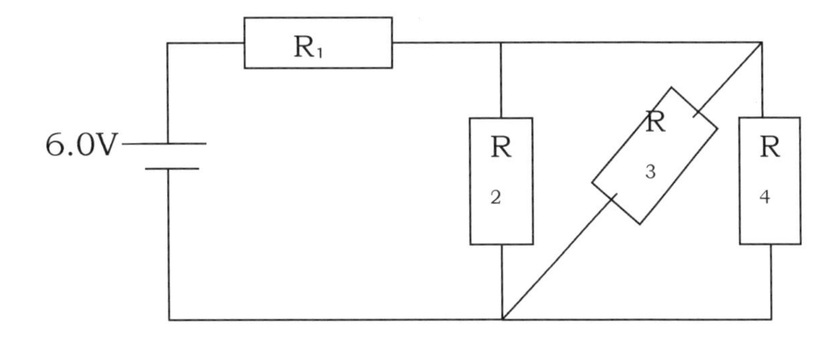

i. What is the combined resistance of the circuit?

ii. What is the current through R_2?

Solution:

a. *Factors affecting the resistance of a conductor:*

- *Length*
- *Thickness*
- *Temperature*
- *Nature of the conductor (material)*

b. i. $\frac{1}{R} = \frac{1}{R_2} + \frac{1}{R_3} + \frac{1}{R_4}$

$= \frac{1}{50} + \frac{1}{50} + \frac{1}{100}$

$= \frac{2 + 2 + 1}{100} = \frac{5}{100}$

$\therefore R = 20\Omega$

$R_T = R_1 + R = 100 + 20 = 120\Omega$

ii. *Total current,* $I_T = \frac{6.0V}{120\Omega} = 0.05A$

P.d across $R_2 = 6 -$ *P.d. across* R_1

$= 6 - (0.05 \times 100) = 1V$

Current through $R_2 = \frac{1V}{50\Omega}$

$= 0.02A$

4. Study the diagram below:

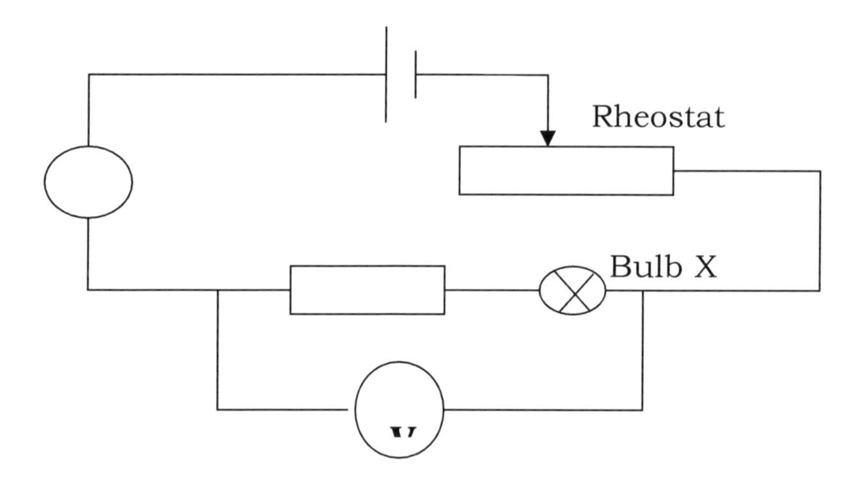

The current I through bulb X is measured for several values of p.d., V across the bulb. The following results were obtained.

V (volts)	1.0	1.5	2.4
I (Amps)	0.20	0.25	0.30

Explain why the resistance is not constant.

Solution:

- *When current increases, the filament gets hot (as a result of collisions of free electrons with atoms, giving some of the kinetic energy to the atoms).*
- *The atoms vibrate faster, impending the flow of electrons more than before. Hence resistance increases.*

5. Four resistors 10Ω, 5Ω, 10Ω and 5Ω are connected to 2.2V supply as shown in the diagram.
Calculate the current through:
a. AD

b. DC

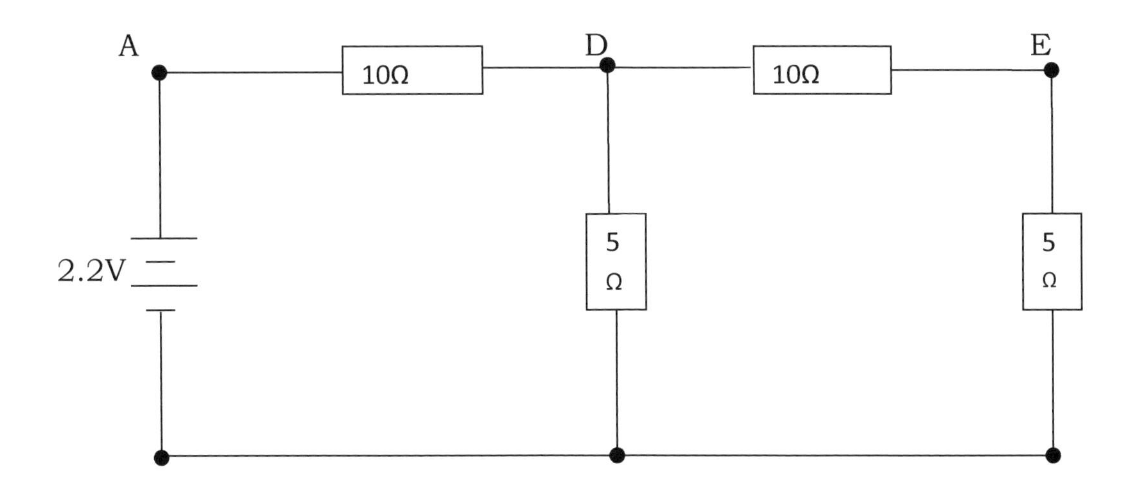

B C F

Solution:

Resistance through DEF = 10 + 5 = 15Ω

DEF is in parallel with 5Ω resistor between D and C.

∴ Combined resistance in parallel is given by

$$\frac{1}{R} = \frac{1}{15} + \frac{1}{5} = \frac{4}{15}$$

$$\Rightarrow R = \frac{15}{4} = 3.75\Omega$$

Thus total resistance, $R_T = 10 + 3.75 = 13.75\Omega$

a. *Current through AD*

$I_{AD} = V/R = 2.2/13.75$

$= 0.16A$

b. *Current through DG*

DEF (15Ω) is in parallel with DG (5Ω)

$$IDG = \frac{15 \times 0.16}{(15 + 5)}$$

= 0.12A (Ratio method)

OR

Since 15Ω and 5Ω are in parallel

Combined resistance $R = \frac{15 \times 5}{15 + 5}$

$= 3.75\Omega$

P.d. across 3.75Ω $V = IR = 0.16 \times 3.75 = 0.6V$

Thus current through 5Ω $I = 0.6/5 = 0.12A$

6. a. Explain what is meant by a pylon as used in mains electricity.

b. Give two advantages of overhead transmissions of electrical power.

c. Give an advantage of three-phase electrical power generation.

d. Explain why electricity is transmitted at high voltage like 400KV rather than at low voltages like 240V.

e. The figure below shows the reading on n electricity meter at the start and end of December.

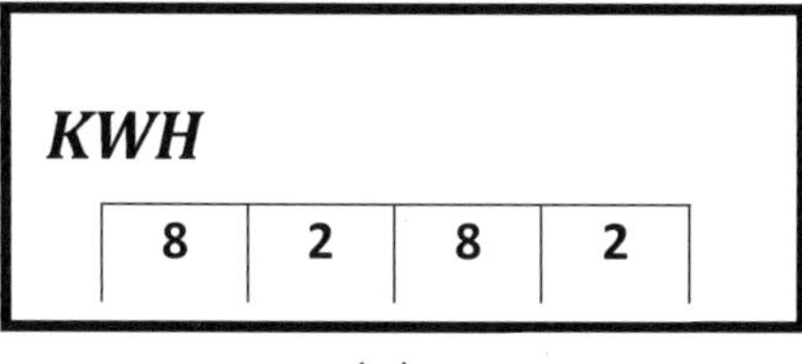

(a)

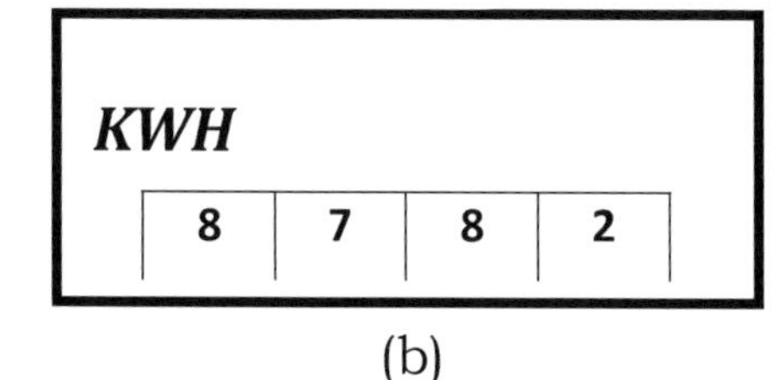

(b)

i. How many KWH have been used during the month of December?

ii. The electrical bill for December was Kshs.750. What was the cost of one electricity unit?

Solution:

a. *Tall post used to suspend electrical cables in overhead transmission system.*

b. - *Electrical energy can be obtained from any location where the power lines pass.*
- *Electrical fault can be easily detected and collected.*
  a. Less expensive
  b. Easily maintained.

c. *The circuit for all the three phases*
  c. Reduce cost of transmission line.
  d. Different consumers can be powered in different phases.
  e. One phase can be put off while the other phase

will continue to supply power.

d. *Very high voltages allow much lower current hence lower power losses as heat.*

e. i. *500kWh*

ii. *Cost of one unit* = ksh. $\frac{750}{500}$

= kshs. 1.50

7. The e.m.f. of a cell is 1.6V. When connected in a circuit whose resistance is 10Ω, the potential difference between its terminals is 1.4V. Calculate the internal resistance of the cell.

Solutions:

e.m.f. = IR + Ir 1.6V = 1.4 + Ir Ir = 1.6 – 1.4 = 0.2	0.14r = 0.2 $r = \frac{0.2}{0.14}$ = 1.43Ω

8. Determine the ammeter reading when a p.d. of 4.5V is applied across PQ in the figure shown below.

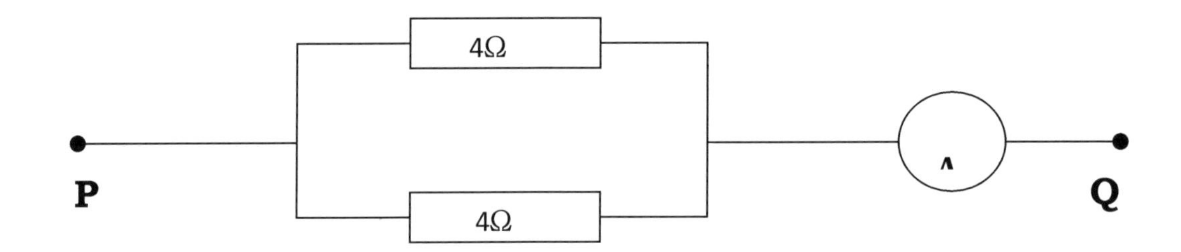

Solution:

$R_T = \frac{4 \times 4}{4 + 4} = \frac{16}{8} = 2\Omega$

V = IR

I = V/R = 4.5/2 = 2.25A

9. In an experiment to determine the forward bias characteristic of a diode,
the following results were obtained.

Voltage (V)	0.0	2.5	3.5	5.0	5.7	5.8
Current (I)	0.0	0.5	1.5	5.0	37.5	70.0

a. Plot a graph of voltage against current.

b. Use the graph to find the breakpoint voltage.

Solution:

a.

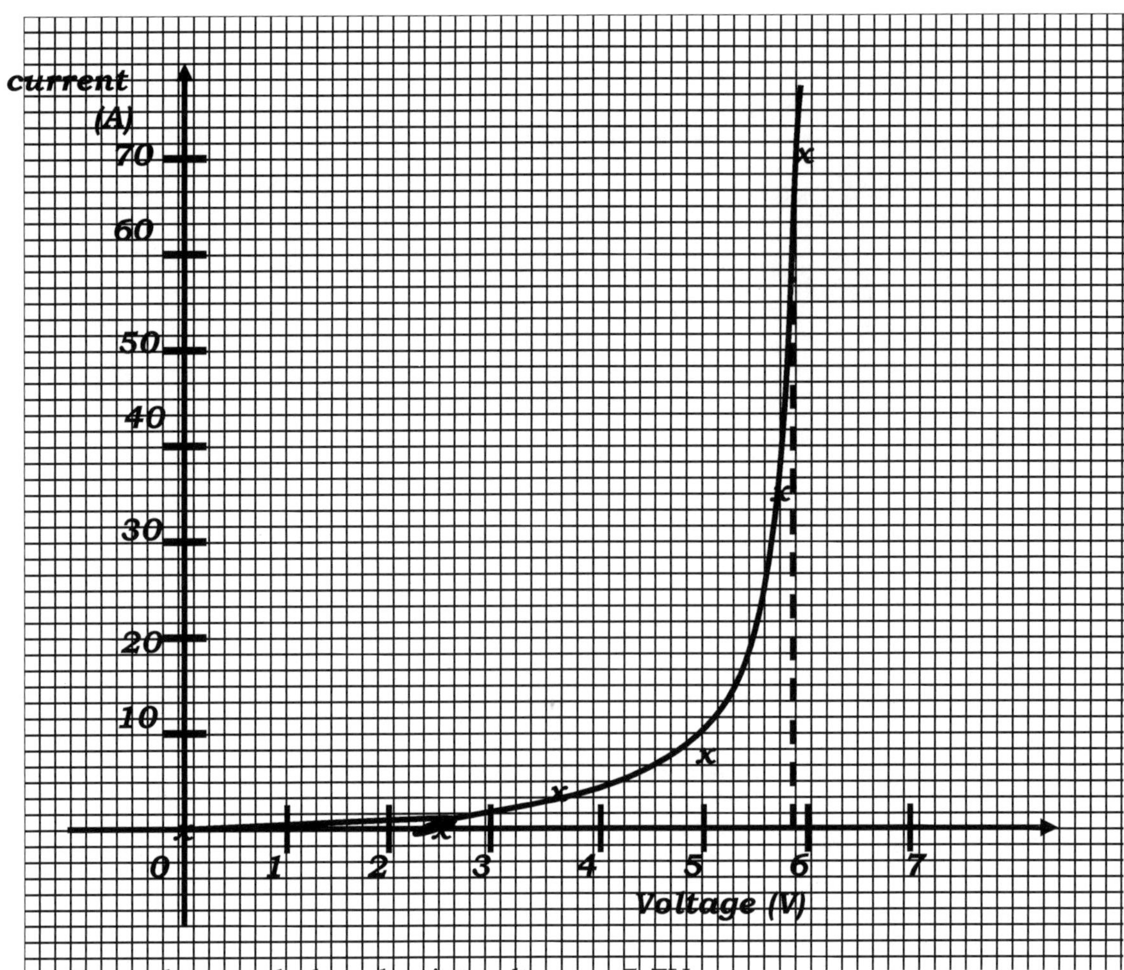

b. *From the graph, breakpoint voltage ≃ 5.7V*

10. a. Explain what is meant by electric rectification.

b. The diagram shows current rectification circuit.

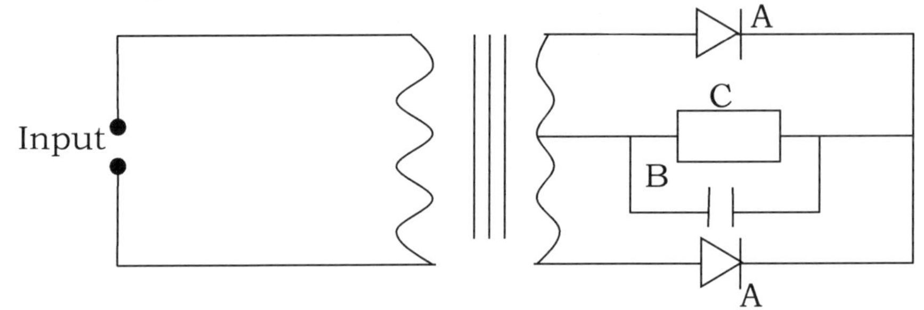

i. Name the component labelled A

ii. Explain the action of component A in causing rectification.

iii. Name the component labelled B.

iv. Explain the use of component B.

v. Suggest a suitable input for voltage for the above circuit.

vi. Name the kind of rectification in the circuit.

> ***Solution:***
>
> a. *This is conversion of alternating current into direct current.*
>
> b. i. *Junction diode*
>   ii. *It allows current to flow in one direction.*
>   iii. *Capacitor*
>   iv. *It smoothens direct current output.*
>   v. *240V – 250V*
>   vi. *Half-rectification*

11. The circuit shown in the diagram is a common-emitter connection used as a switch.

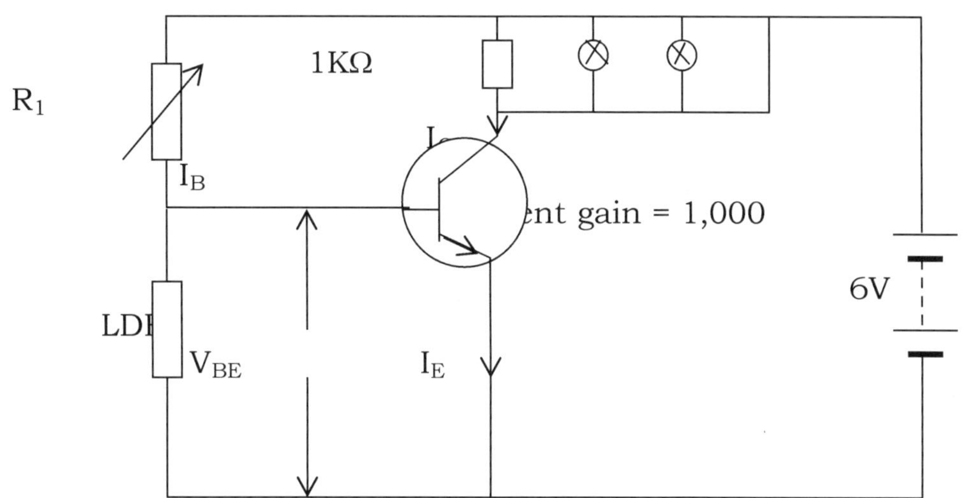

a. Explain how the transistor switches on the bulbs in the output circuit work.

b. Suggest a suitable application for the circuit.

c. If all the bulbs are switched on V_{CE} = 2V, what is the p.d. across the bulbs

d. If current I_C is 1.0 x 10^{-2}A, determine the value of I_B.

e. Obtain the value of I_E.

Solution:

a. *When LDR is covered, the resistance rises and voltage increases switching on the transistor.*

b. *Switching on streetlights when the streets become dark.*

c. $V_{CC} = I_C R_C + V_{CE}$

$\Rightarrow \quad 6V = I_C R_C + 2.0V$

$\Leftrightarrow \quad I_C V_C = 4V$

d. $h_{FE} = \frac{I_C}{I_B} \quad \Rightarrow \quad 1000 = \frac{1.2 \times 10^{-2}A}{I_B}$

$\therefore \quad I_B = \frac{1.2 \times 10^{-2}A}{1000} = 1.2 \times 10^{-5}A$

e. $I = I_C + I_B$

$\Leftrightarrow \quad I = 1.2 X 10^{-2}A + 1.2 X 10^{-5}$

$= 1.212 X 10^{-3}A$

12. The element of an electric hot plate has a resistance of 40Ω. What is the energy dissipated when the element is on for 20 minutes on a 240V supply?

Solution:

$V = IR \quad \Leftrightarrow \quad I = \frac{V}{R} = \frac{240}{60} = 4A$

$Energy = Pt = IVt$

$= 4 \times 240 \times 20 \times 60 = 1{,}152{,}000J$

$= 1.152 \times 10^6 J$

13. The diagram shows a network of resistors and the current flow.

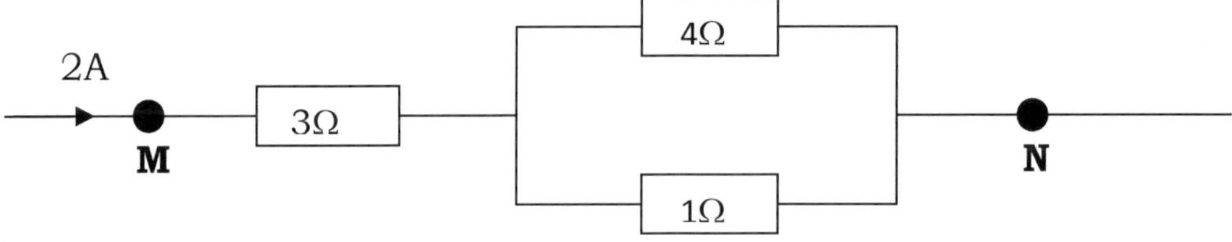

Determine:

a. The p.d. across the points MN.

b. The current through the 1Ω resistor.

Solution:

a. Resistance in parallel $= \frac{4 \times 1}{4 + 1} = \frac{4}{5} = 0.8\Omega$

$\therefore$ Total resistance $= 3 + 0.8 = 3.8\Omega$

$V = IR$

$= 3.8 \times 2 = 7.6V$

b. p.d. across 3Ω resistor $= 3 \times 2 = 6V$

$\therefore$ p.d across resistors in parallel $= 7.6 - 6 = 1.6V$

$\Rightarrow$ Current trough the 1 resistor:

$I = \frac{V}{R} = \frac{1.6}{1} = 1.6A$

14. Four-40W bulbs and seven-100W are switched on for 3 hours a day for domestic lighting. Find the monthly bill for the consumer given that the cost of electricity is at shs. 3.50 per unit. (Take 1 month = 30 days.)

Solution:

Power consumed per month $= \frac{(4 \times 40 + 7 \times 100) \times 3 \times 30}{1{,}000}$

$= \frac{860 \times 90}{1{,}000} = 77.4kWh$

$\therefore$ Cost per month $= 77.4 \times 3.50 =$ shs.270.90

15. In a p-type semi conductor, holes are the main charge carriers. Explain.

Solution:

As the group three element fit itself in the group four element, there is a vacant site that is created.

16. A transformer in a welding machine supplies 6 volts from a 240V mains supply. If the current used in the welding is 30A, determine the current in the mains.

Solution:

Power input = power output

$\Rightarrow \quad 240I = 30 \times 6$

$\Leftrightarrow \quad I = \frac{180}{240} = 0.75A$

17. The figure shows an electric circuit consisting of a battery of e.m.f. 3 volts and an internal resistance of 1.3 ohms, joined to two resistors of 2 ohms and 3 ohms connected in parallel.

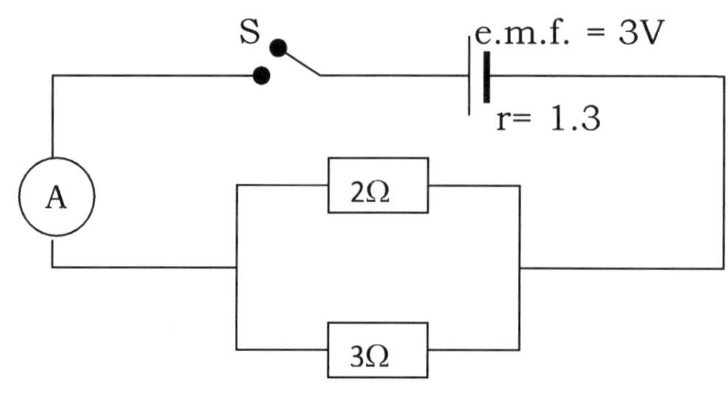

Calculate the ammeter reading when the switch 'S' is closed.

Solution:

$I = \frac{V}{R}$ *Where R is the combined resistance.*

$R = \frac{2 \times 3}{2 + 3} + 1.3 = 1.2 + 1.3$

$R = 2.5\Omega \quad \therefore \quad I = \frac{3}{2.5} = 1.2A$

18. A fully charged battery carried a mark "100A" on it. A radio is connected to it and draws a current of 2.5A. If the radio is on 2 hours daily, how many days will it take before the battery is due for charging?

Solution:

$$Q = It \Rightarrow 100Ah = 2.5 \times t$$

$$\therefore \quad t = \frac{100Ah}{2.5} = 40hrs$$

19. Name two advantages, which a lead-acid accumulator has over a dry cell.

Solution:

- *High current*
- *High voltage*

20. Explain how the earth wire, along with the fuse fitted to the plug guard against electrocution.

Solution:

- *If a fault occurs so that the casing becomes live, a large current passes to the earth.*
- *The fuse melts breaking the connection in the wire and the casing is no longer live.*

21.

Chapter three: Density, Relative Density and Pressure

1. The figure represents a block of uniform cross-sectional area of $6.0cm^2$ floating on two liquids A and B. The lengths of the block in each liquid are shown.

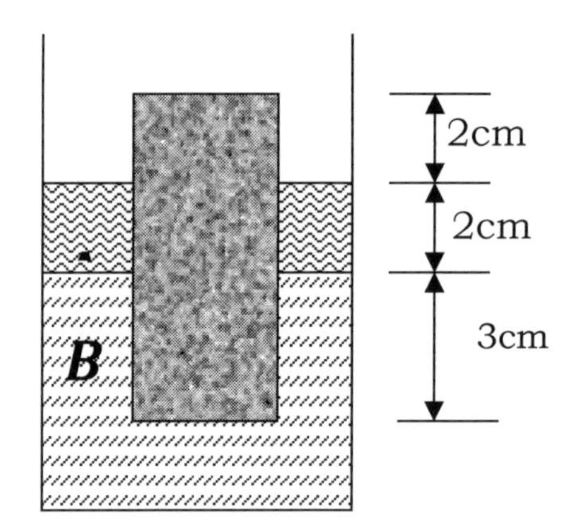

Given that the densities of liquids A and B are $800kgm^{-3}$ and $1,000kgm^{-3}$ respectively, determine the:

a. Weight of liquid A displaced.

b. Weight of liquid B displaced.

c. Density of the block.

Solution:

a. *Weight* = *mass x g*

Mass = *density x volume*

$= 800 \times \frac{6}{10,000} \times \frac{2.}{100}$

$\rightarrow$ *Weight* $= 800 \times 1.2 \times 10^{-5} \times 10$

$= 0.096N$

b. *Density of block* $= \frac{\text{mass of block.}}{\text{Volume of block}}$

$= \frac{(0.0\,96N + 0.18N)kg}{10}$

$= 0.0276kg.$

c. *Volume of block* $= \frac{6 \times 7}{1,000,000} m^3 = \frac{42}{1,000,000}$

$\rightarrow$ *Density* $= \frac{0.0276 \times 1,000,000}{42}$

$= 657.4kg/m^3$

2. a. The table shows values of pressure, P in fresh water at different depths, h.

Pressure, p (KPa)	110	140	180	200	220
Depth, h (m)	1.0	4.0	8.0	10.0	12.2

i. Plot a graph of pressure (y-axis) against depth.

ii. Given the equation $P = P_0 + h\rho g$ determine from the graph the value of P_0.

iii. Determine the density of the fresh water.

b. In an experiment to determine the relative density of a solid block, the following measurements were obtained.

- Weight of block in air = 0.237N
- Weight of block when immersed in water = 0.207N

Use the measurements to determine the density of the solid. (Density of water = $1gcm^{-3}$)

Solutions:

a. *i.* *See the graph attached.*

ii. *From the equation, P_o = y-intercept = 100Kpa.*

iii. *From the equation, ρg = gradient*

Gradient of the graph = 10

$10g = 10$

$g = 1g/cm^3$.

b. *Upthrust = wt in air – wt in water*

= 0.237 – 0.207 = 0.03N

Relative density $= \frac{\text{wt in air}}{\text{Upthrust}} = \frac{0.237}{0.03} = 7.9$

Density of the solid = 7.9 x 1

= $7.9g/cm^3$

3. a. State the law of floatation.

b. A test-tube, cross sectional area $6.4cm^2$ is loaded with lead shots so that it can be used as a hydrometer, as shown in the diagram.

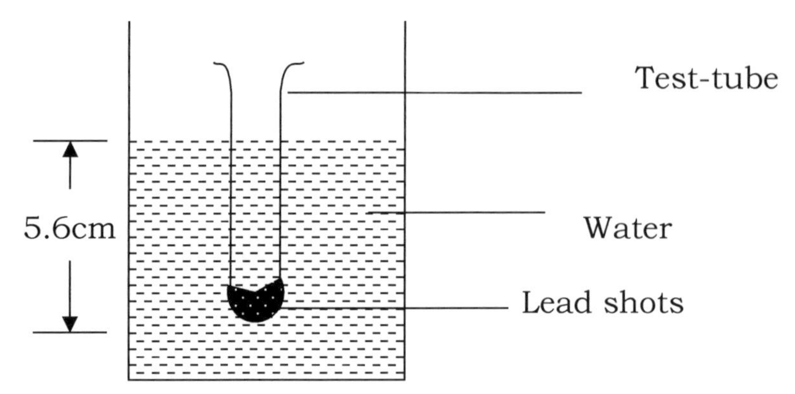

i. What is the mass of the hydrometer given that density of water = $1gm^{-3}$?

ii. When put in liquid X, the test tube sinks to a depth of 4.8cm. What is the density of liquid X?

Solution:

a. *A floating object displaces its own weight in a fluid in which it floats.*

b. i. *Volume of water displaced*
$= (6.4 \times 5.6)cm^3$
$= 35.84cm^3$
Mass of water displaced = mass of the hydrometer
$= 35.84 \times 1$
$= 35.84g.$

ii. $\frac{d_1}{d_2} = \frac{h_2}{h_1}$ $h_1 = 5.6cm$
$h_2 = 4.8cm$

$\frac{d_1}{1} = \frac{5.6}{4.8}$

$d_1 = 1.167g/cm^3$

4. A metal block of density 5,000kg/m^3 measuring 80mm by 40mm by 25mm is attached to a spring balance A as shown in the diagram. The mass of an empty beaker on the compression balance B is 100g,

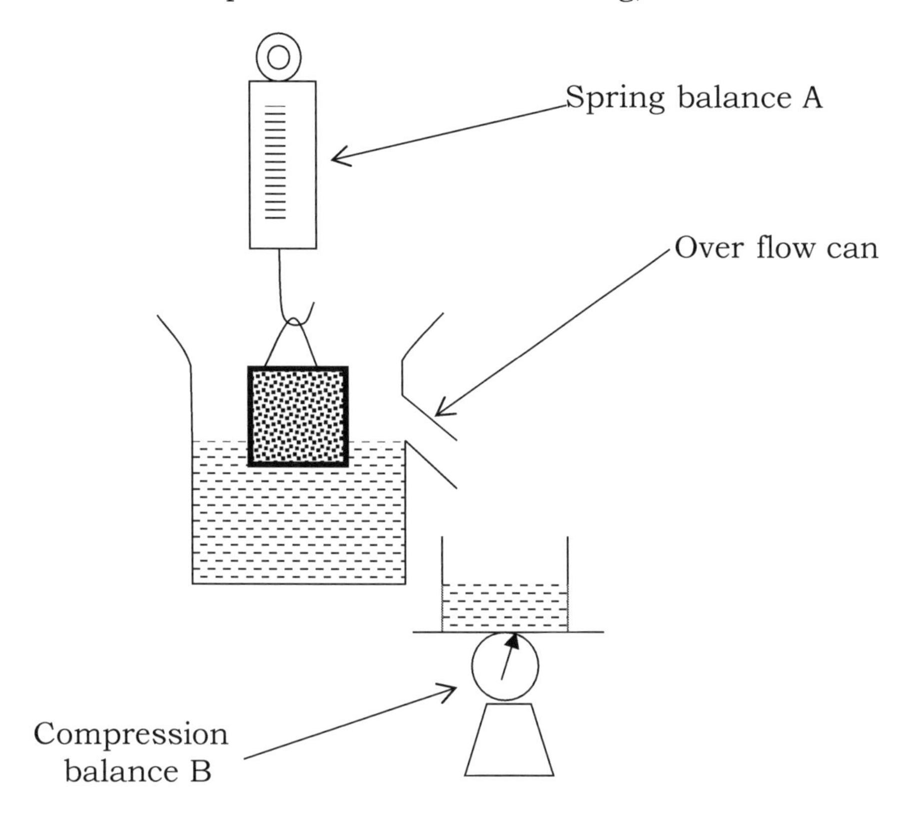

a. What are the readings of the spring balance A and the compression balance B when:

   i. The metal block is in air.

   ii. The block is lowered slowly until it is partially submerged to a depth of 10mm.

   iii. The block is now lowered such that it is completely just below the surface of water.

   iv. The block is now lowered until it rests on the bottom of the beaker.

b. Draw a diagram of the block showing all the forces acting on it when it is completely submerged in water.

c. If the metal block was completely immersed in a liquid of density 800kg/m^3:

i. Will the spring balance read more or less than it read when the block was immersed in water? Give reason for you answer.

ii. What will A read?

d. The experiment is repeated with a block of wood of the same dimension and density 800kg/m^3.

i. What is the initial reading in A?

ii. What happens to the block when it is lowered into the water?

Solution:

a. i. Mass of the block = density x volume

= 500 x 0.08 x 0.04 x 0.025

= 0.4kg

∴ Weight of the block = 4N

⇔ A reads 4N while B reads 100g.

ii. As the block is immersed, water is displaced into the beaker so the reading on B increases.

∴

Volume of block immersed = 8.0 x 4.0 x 1.0cm^3

= 32cm^3

⇒ Volume of water displaced = 32cm^3

Mass of water displaced = 32g

(Density of water = 1g/cm^3)

Upthrust of water = 0.32N

∴ The spring balance reads 4.00N – 0.32N

= 3.68N

Balance B reads 100g + 32g = 132g

iii. When the block is completely just under the surface of water, volume of water displaced = volume of the block

= 8.0 x 4.0 x 2.5 = 80cm^3

Mass of water displaced = 80g

∴ Upthrust of water = 0.8N

⇔ The spring balance reads 4.0N – 0.8N = 3.2N and balance B reads 100g + 80g = 180g

As the block is lowered further, no more water is displaced since it is already completely submerged.

∴ The readings on A and B remain steady until

the block rests on the bottom of the can.

iv. *The spring balance reads 0(zero) while B reads 180g.*

b.

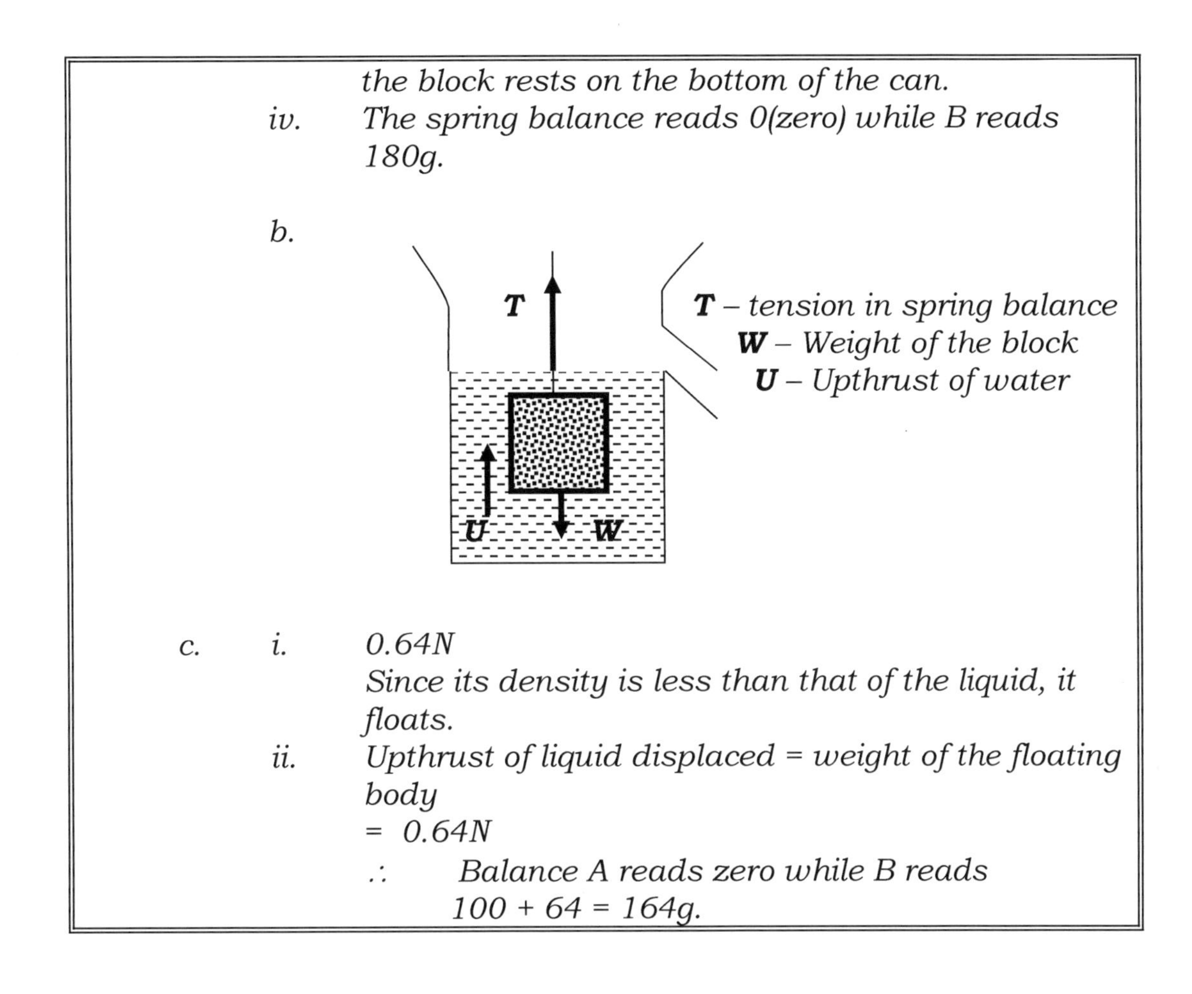

c. i. *0.64N*

Since its density is less than that of the liquid, it floats.

ii. *Upthrust of liquid displaced = weight of the floating body*

= 0.64N

∴ Balance A reads zero while B reads 100 + 64 = 164g.

5. If the atmospheric pressure is 760mmHg, calculate the pressure in Pascal's of the trapped air, x, in the tube shown in the figure. (Take density of mercury = 13.6g/cm^3)

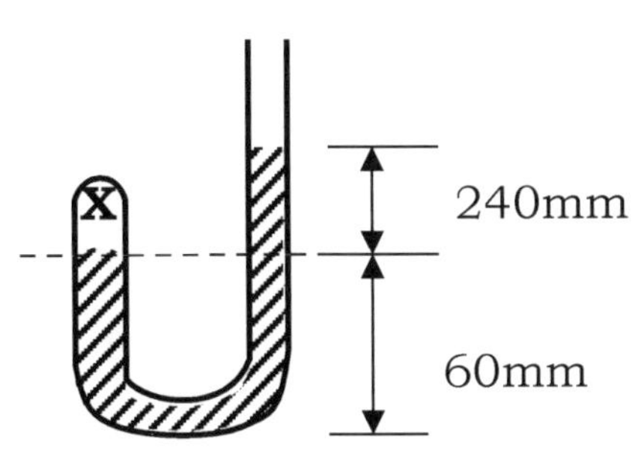

Solution:

Pressure of x, P = $h\rho g$ + atm. Pressure

= 13,600 x 10(0.24 + 0.76)

= 136,000Pa

6. a. Define pressure
(1 mark)

b. A hole is made at the middle of a tall container and water is filled in the it as shown in the diagram.

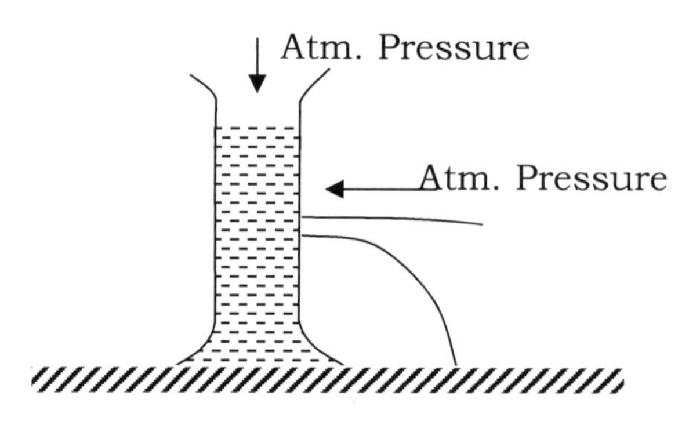

Explain:

- Why does the water come out of the hole and yet the atmospheric pressure is acting in the directions shown.
- What would be observed if the container was filled with water and the top open part was closed very tightly?

Solution:

a. Pressure is force acting perpendicularly per unit surface area.

b. The height 'h' of the water column above the hole provides the pressure ($P = h\rho g$), which makes the water to come out. Closing the top, it will cut off the atmospheric pressure from top hence making the atmospheric pressure acting at the hole to be greater than the pressure inside due to the water column.

7. A cube of wood of side 5cm and density $600Kg/m^3$ is placed in water. What force must be applied to the cube so that the top surface of the cube is on the same level as the water surface? (Density of water = $1,000Kg/m^3$ and g = 10N/Kg)

Solution:

Volume of the cube = $(0.05)^3 m^3$
Weight of the cube = $(0.05)^3 \times 600 \times 10N$
Upthrust = *Weight of water displaced*
= $(0.05)^3 \times 1{,}000 \times 10N$

$F = U - W$
$= [(0.05)^3 \times 1{,}000 \times 10] - [(0.05)^3 \times 600 \times 10]$
$= (0.05)^3 \times 400 \times 10$
$= 0.5N$

8. The pressure of the liquid column AB supports a mass M on piston C.

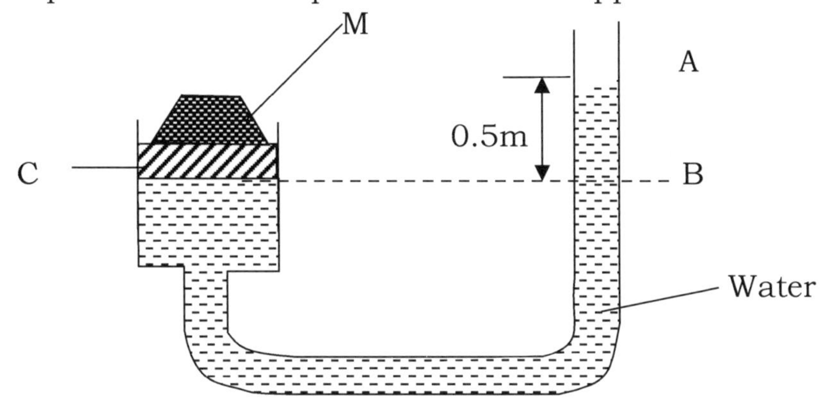

If the area of piston C is $0.1m^2$, calculate the value of M. (Take density of water = $1{,}000Kg/m^3$)

Solution:

Pressure = $h\rho g = F/A$
$\Leftrightarrow$ $F = h\rho gA$
$= 0.5 \times 1{,}000 \times 10 \times 0.1$
$= 500N$
$\therefore$ *Mass m* = $\frac{500}{10} = 50Kg$

9. A body weighs 32N in kerosene and 30N in water. If the body weighs 40N in air, what is the density of kerosene? (Density of water = $1g/cm^3$)

Solution:

$$\text{Relative density} = \frac{\text{Density of kerosene}}{\text{Density of water}}$$

$$= \frac{\text{Weight of kerosene displaced}}{\text{Weight of equal volume of water}}$$

$$= \frac{40N - 32N}{40N - 30N} = 0.8$$

$$\therefore \quad \text{Density of kerosene} = 0.8 \times 1g/cm^3$$

$$= 0.8g/cm^3$$

Solution:

$$\text{Relative density} = \frac{\text{Density of kerosene}}{\text{Density of water}}$$

$$= \frac{\text{Weight of kerosene displaced}}{\text{Weight of equal volume of water}}$$

$$= \frac{40N - 32N}{40N - 30N} = 0.8$$

$$\therefore \quad \text{Density of kerosene} = 0.8 \times 1g/cm^3$$

$$= 0.8g/cm^3$$

Chapter four: Electrostatics and capacitance

1. Two capacitors of 3μF and 6μF respectively are arranged in series with a battery e.m.f. 100V as shown in the figure.

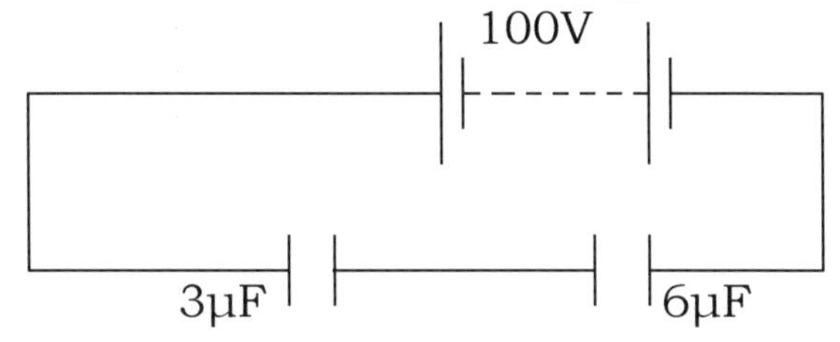

Determine:

a. Effective capacitance for the arrangements.

b. The total charge stored in the arrangements.

c. The total energy stored in the capacitors.

Solutions:

a. *Effective capacitance* $C = \frac{6 \times 3}{6 + 3} = 2\mu F$

b. $Q = CV = 2^2 \times 10^{-6} \times 100 = 2 \times 10^{-4} C$

c. *Total energy* $= \frac{1}{2}CV^2 = \frac{1}{2} \times 2 \times 10^{-6} \times 100^2$

$= 1 \times 10^{-2}$ *joules*

OR

$E = \frac{1}{2}QV = \frac{1}{2} \times 2 \times 10^{-4} \times 100 = 1 \times 10^{-2}$ *Joules*

2. a. State three factors that affect the capacitance of a capacitor.

b. In the figure below, calculate the charge stored in the 2μF capacitor.

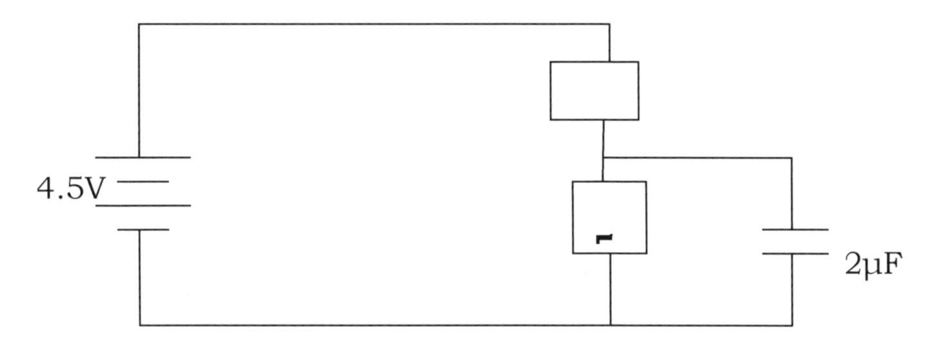

Solution:

a. - *Area of plates*

- *Distance of separation of plates*
- *Presence of dielectric material*

b. $I_T = \frac{4.5V}{15\Omega} = 0.3A$

p.d. thro" 2μF = p.d thro' 5Ω = 0.5 x 5 = 1.5V

$Q = CV$

$= 2 \times 10^{-6} \times 1.5V$

$= 3.0 \times 10^{-6}$ *coulombs*

3. A negatively charged rod is brought near the cap of a leaf electroscope. The cap is then earthed by touching a finger. Finally the rod is withdrawn. The electroscope is found to be positively charged. Explain how the positive charge is acquired.

Solution:

- *On bringing the negative charged rod near the cap, negative charges are repelled from the cap to the leaf and the plate.*
- *On earthing negative, charges from the leaf and plate flow through the finger thus neutralising them.*
- *On removing the changing and the finger, the positive charges on the cap attracts negative charges from the leave and the plate, making them positive charged.*

4. a. Define capacitance.

b. A 12μF capacitor is charged with a 200V source and then placed in parallel with an 8μF. Calculate

i. The initial charge on the 12μF capacitor.

ii. The final p.d. across each capacitor

iii. The final charge on each capacitor.

Solution:

a. *A measure of the extent to which a capacitor can store charge or charge storing ability of a capacitor.*

b. i. $Q = VC$

$= 200 \times 12 \times 10^{-6} = 2.4 \times 10^{-3} C$

ii. *Total capacitance* $= 12 + 8 = 20\mu F$

$V = \frac{Q}{C} = \frac{2.4 \times 10^{-3}}{20 \times 10^{-6}} = 120V$

iii. *For* $12\mu F \Rightarrow Q = 120 \times 12 \times 10^{-6}$

$= 1.44 \times 10^{-3} C$

For $8\mu F \Rightarrow Q = 120 \times 8 \times 10^{-6}$

$= 9.6 \times 10^{-4}$

5. Study the circuit shown in the diagram.

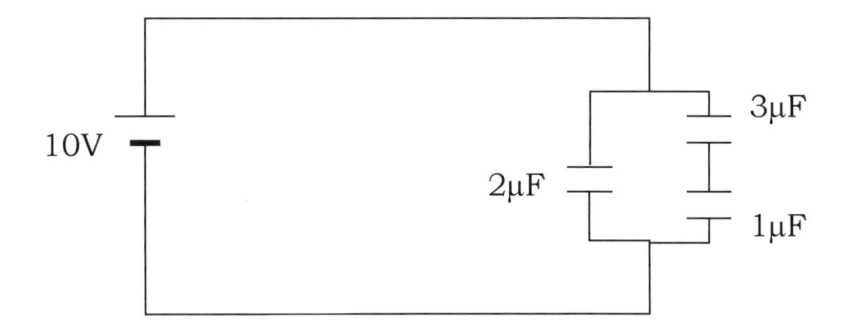

a. Determine the charge stored in the 2μF capacitor.

b. What is the combined capacitance of the arrangement?

Solution:

a. $Q = CV$

$= 2 \times 10^{-6} \times 10$

$= 2.0 \times 10^{-5} C$

b. *Capacitance in series:*

$= \frac{3 \times 1}{3 + 1} = \frac{3}{4} = 0.75\mu F$

$\therefore$ *Total capacitance*

$= 0.75 + 2 = 2.75\mu F$

6. Explain why an uncharged metal rod brought close to but not touching the cap of a charged electroscope causes a decrease in the divergence of the leaf.

Solution:

The uncharged rod has both charges, hence the unlike charges attract causing the leave to fall.

7. During the process of charging an electroscope by induction, a positively charged rod is brought close (but not touching) te cap of an electroscope. The leave diverges. State and explain the observations when the cap is earthed with the rod still in position.

Solution:

- *The leaf collapses.*
- *Electrons from the earth neutralise the positive charges on the leaf.*

8. State three uses of a gold leaf electroscope.

Solution:

- *To detect the presence of charge on a body.*
- *To test the type of charge on a body*
- *To test the quantity of charge on a body.*
- *To test for insulation properties of materials.*

Chapter five: Energy and Power

1. The data given below shows experiment results of potential energy obtained when a stone was dropped from different heights above the ground.

h (m)	0	1	2.2	4.2	5.6	7.4	9.2
p.e x 100(J)	0	0.7	1.3	2.4	3.1	3.9	4.7

a. i. Plot a graph of p.e. against height, h.

ii. Calculate the slope of the gradient.

i. What does the slope signify?

ii. Use the results in (ii) above to estimate the mass of the stone used.

iii. Give the energy changes that take place as the stone falls on the ground.

b. Calculate the velocity with which a table tennis ball of mass 10g leaves the bat when a person uses energy of 200J.

2. A pump uses a mixture of petrol and alcohol in the ratio 4:1 by mass, to pump water from a well 200m deep. (Energy per gram of alcohol = 27×10^3 J/g, and petrol = 48×10^3 J/g)

a. How much energy is given by one gram of the mixture.

b. If the pump is 60% efficient, what mass of this mixture is needed to raise the water?

Solution:

a. *1 gram of mixture has 4/5 gram of petrol*

$= 4/5 \times 48 \times 10^3 J = 38{,}400J$

and 1/5 gram of alcohol $= 1/5 \times 27 \times 103J = 5{,}400J$

$\therefore$ *Total energy* $= 38{,}400 + 5{,}400 = 43{,}800J/g$

b. *Energy to raise water = mgh*

$= 1{,}000kg \times 10 \times 200m = 2 \times 10^6 J$

Energy used per gram $= 60\% \times 43{,}800 = 26{,}280J/g$

Mass of water to raise $= \frac{2 \times 10^6}{26{,}280}$

$= 76.10g$

3. a. A block and tackle system has three pulley wheels in each block. It was used to perform an experiment and the results that were obtained were as follows.

Load (N)	25	75	130	180	230
Effort (N)	10	20	30	40	50
Mechanical advantage					
Efficiency %					

i. Complete the table.

ii. Draw a graph of efficiency against the load.

b. From the graph,

i. Give the reasons why the efficiency of the system can never be 100%.

ii. Calculate the effort when the efficiency is 70%

Solution:

a. *i.*

Load (N)	*25*	*75*	*130*	*180*	*230*
Effort (N)	*10*	*20*	*30*	*40*	*50*
Mechanical advantage	***2.5***	***3.75***	***4.33***	***4.5***	***4.6***
Efficiency %	***41.7***	***62.5***	***72.2***	***75***	***76.7***

ii.

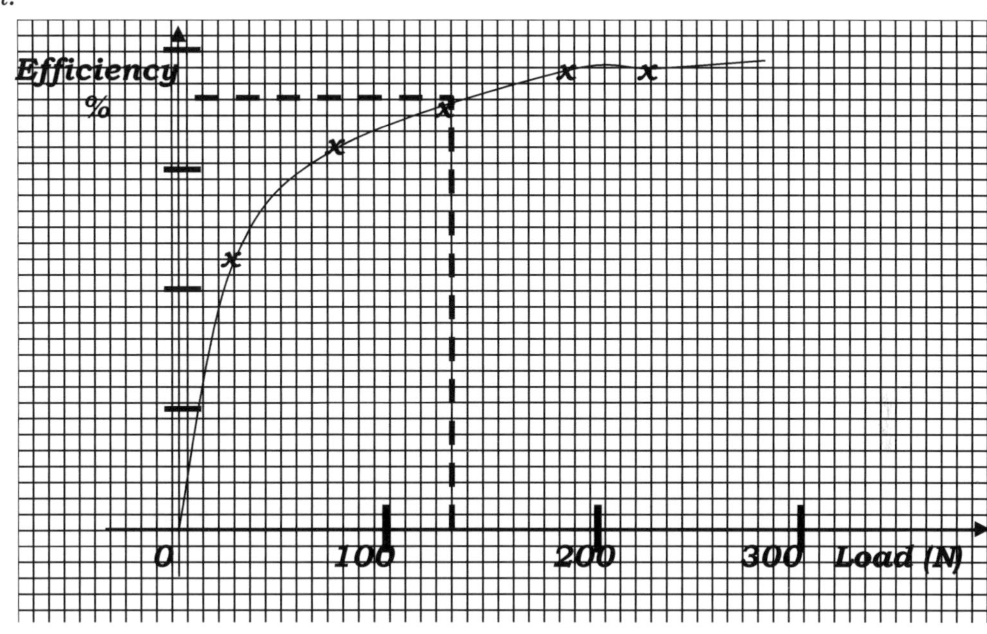

b. *i.* - *Extra effort is applied to overcome friction between groove and tread.*

- *Extra effort is applied to overcome weight of movable pulley.*

ii. *At 70% efficiency, load ≃ 130N*

$$\Leftrightarrow Effort = \frac{Load \times 100}{Efficiency \times VR}$$

$$= \frac{130 \times 100}{70 \times 6}$$

$$\simeq 31N$$

4. a. Distinguish between potential and kinetic energy.

b. A block of mass 5kg is placed atop a 4m high frictionless inclined plane as shown in the figure below.

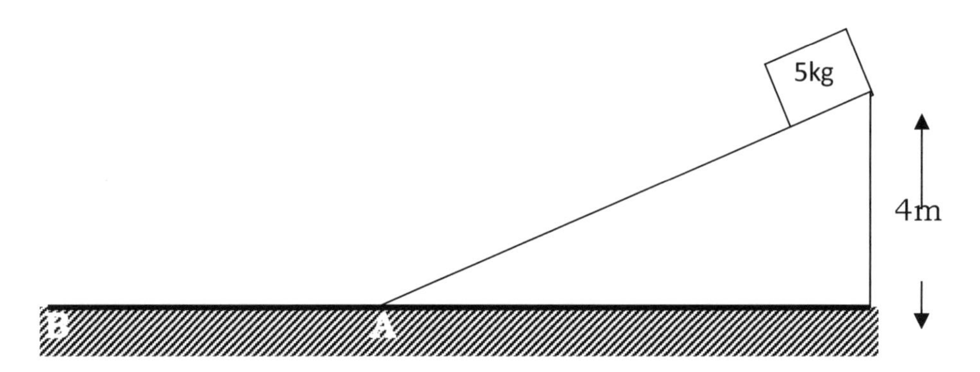

i. Determine the potential energy of the block at its position.

ii. If the block slides freely down the incline to a point A, then decelerates uniformly to a stop at point B, calculate the distance AB given that the coefficient of friction is 0.2.

c. A student determined the potential energy of an object M on its way up a staircase as shown in the figure below. She then tabulated her results as shown in the table.

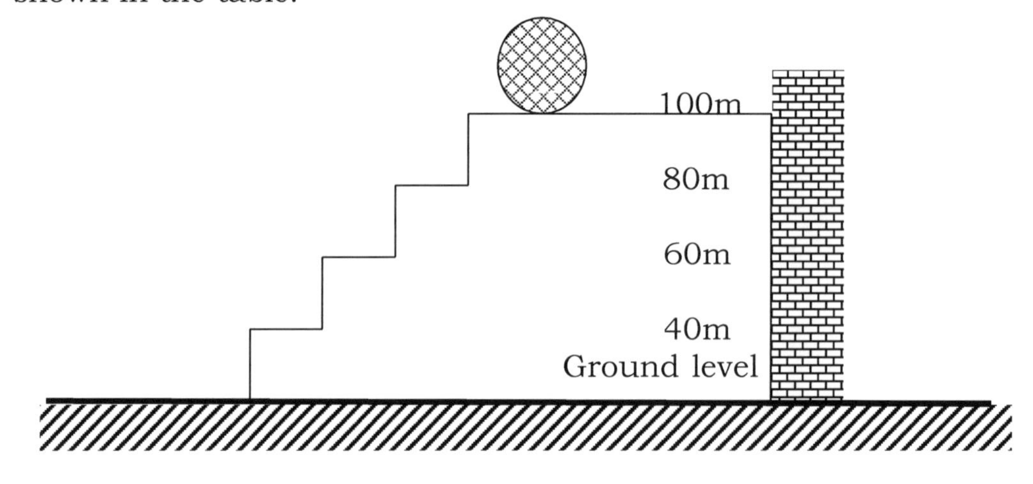

Height (m)	0	20	40	60	80	100
Potential energy (J)	0	400	800	1200	1600	2000
Kinetic energy (J)						

i. If the object is allowed to fall back freely (vertically), fill the kinetic energy row to show how it changes with the height.

ii. On the same set of axes, plot the graphs of P.E (vertical axes) against height and K.E (vertical axis) against height, labelling clearly.

iii. What physical quantity is represented by the gradient of the P.E.-height graph?

iv. At what height does the object possess equal amounts of both energies?

Solution:

a. *Potential energy is the energy possessed by a body due to its position in a field of force while kinetic energy is energy possessed by a body in motion.*

b. *i.* $P.e. = mgh = 5 \times 10 \times 4$

$= 200J$

ii. *Let the distance AB be L and frictional force be F*

$F = \mu R$ *and* $R = mg \therefore F = \mu mg$

Work done to stop the block = potential energy at the top.

$\therefore FL = mgh \Leftrightarrow \quad L = \frac{mgh}{F} = \frac{mgh}{\mu mg} = \frac{h}{\mu}$

$\therefore L = \frac{4}{0.2} = 20m$

5. A crane on a building site is used to lift materials from the ground to the top of a new building. This of materials weighs 10,000N and is lifted 25m at a steady speed.

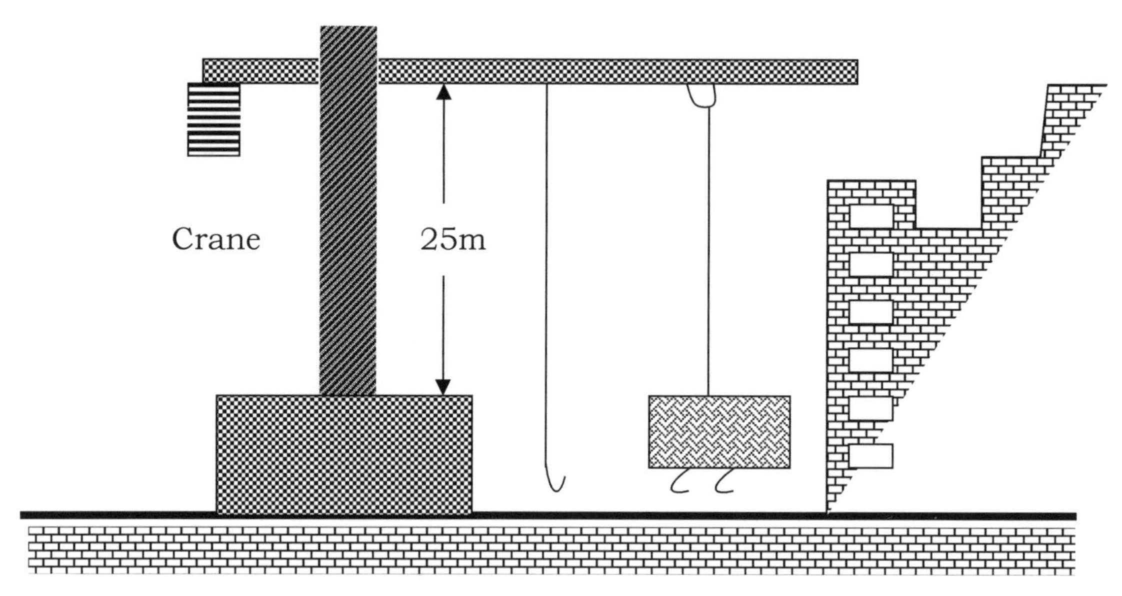

a. Calculate the energy used in lifting the material to the top of the building.

b. What is the power produced by the crane if the lift took 50 seconds.

c. If the electrical energy was actually supplied to the crane at a power of 6kW, calculate the efficiency of the crane.

Solution:

a. *Energy used = work done = mgh*

$$= 10{,}000 \times 25 = 250{,}000 J$$

b. $Power = \frac{Energy}{time} = \frac{250{,}000}{50}$

$$= 5{,}000 \text{ watts}$$

c. $Efficiency = \frac{Work\ output}{work\ input} \times 100\%$

$$= \frac{5{,}000}{6{,}000} \times 100\%$$

$$= 83.33\%$$

6. The diagram represents a pulley system supporting a load of 5kg.

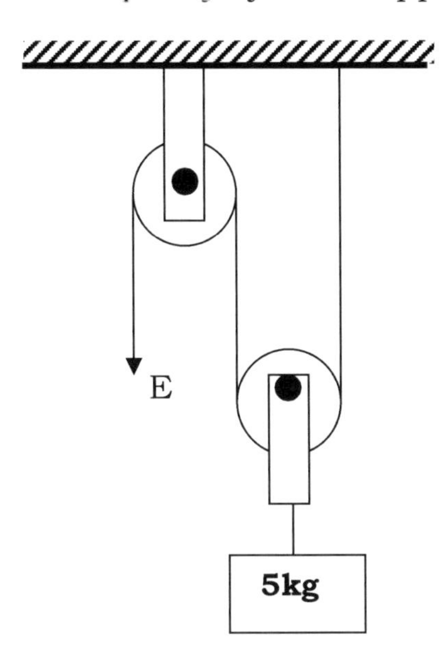

Given that the efficiency of the system is 80%, determine the effort E. (Take g = 10N/kg)

Solution:

$$Efficiency = \frac{Work\ output}{Work\ input} \times 100\%$$

$$= \frac{M.A.}{V.R.} \times 100\%$$

But V.R. of toe system = 2

$$\Rightarrow \quad 80\% = \frac{M.A.}{2} \times 100\% \quad \Leftrightarrow \quad M.A. = \frac{80 \times 2}{100}$$

$$M.A. = 1.6$$

$$And \quad M.A. = \frac{Load}{Effort} \quad \Leftrightarrow \quad Effort = \frac{5 \times 10}{1.6}$$

$$Effort = 31.25N$$

7. An object dropped from a height h, attains a velocity of $6ms^{-1}$ just before hitting the ground. Calculate the value of h.

Solution:

Potential energy lost = kinetic energy gained

$\Leftrightarrow \quad mgh = {}^{1}/{}_{2}mv^2 \quad \Rightarrow \quad v^2 = 2gh$

$$h = \frac{v^2}{2g} = \frac{36}{20} = 1.8m$$

8. The engine of a crane develops a total output power of 20kW. At what speed does it raise a container of mass 1,000kg?

Solution:

$$Power = \frac{Work\ done}{Time\ taken} \quad \Rightarrow \quad P = \frac{Fd}{t}$$

$$Power = force \times velocity \quad \therefore \quad v = \frac{20{,}000W}{1{,}000 \times 10N}$$

$v = 2m/s$

9. A screw jack has a velocity ratio of 314 and an efficiency of 40%. An effort of 25N is exerted on the handle of the jack. Determine the maximum load that can be raised by the jack.

Solution:

$$Efficiency = \frac{M.A.}{V.R} \times 100\% \quad \Rightarrow \quad 40\% = \frac{M.A}{314} \times 100\%$$

$$\Leftrightarrow \quad M.A. = \frac{40 \times 314}{100} = 125.6$$

$$M.A. = \frac{Load}{Effort} \quad \Rightarrow \quad 125.6 = \frac{Load}{25}$$

$\Leftrightarrow \quad Load = 125.6 \times 25 = 3{,}140N$

10. A crane lifts a load of mass 1,000kg at constant speed through a vertical distance of 3.0m in 6.0s. Find the power of the crane. (Take g = 10.0N/Kg)

Solution:

$$Power = \frac{Work\ done}{Time\ taken}$$

$$= \frac{1{,}000 \times 10 \times 3}{6}$$

$$= 50{,}000W$$

11. A machine has a velocity ratio of 4 and efficiency of 80%. Calculate the effort needed to raise a load of 100N using the machine.

Solution:

$$\text{Efficiency} = \frac{M.A.}{V.R.} \times 100\% = \frac{\text{Load}}{\text{Effort}} \times \frac{1}{V.R} \times 100\% = 80\%$$

$$\Rightarrow \quad \frac{100}{E} \times \frac{1}{4} \times 100 = 80 \quad \Leftrightarrow \quad E = \frac{100 \times 100}{4 \times 80}$$

$$\therefore \quad E = 3.125N$$

12. The figure shows a pendulum bob in two positions A and B. The pendulum bob has a mass of 0.3Kg.

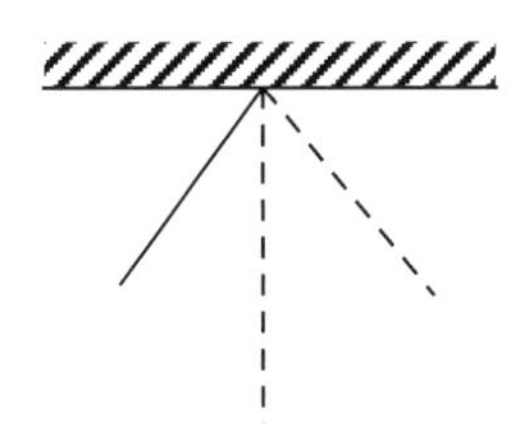

The bob is released from position A and swings to and fro. Find the maximum speed of the bob.

A ○ C

0.8m ○B

Solution

$mgh = \frac{1}{2}mv^2 \quad \Leftrightarrow v^2 = 2gh$

$\therefore \; v^2 = 2 \times 10 \times 0.8 = 16$

$\Leftrightarrow v = \sqrt{16} = 4m/s$

Chapter six: Equilibrium and Centre of Gravity

1. A man carrying two similar buckets full of water on each hand is more stable than the one carrying one such bucket on one hand. Explain why this is so.

> ***Solution:***
> *When carrying two buckets, the man's C.O.G. is somewhere between the legs (base) but when carrying one bucket, the C.O.G is towards one side.*

2. State two factors that reduce the stability of vehicle.

> ***Solution:***
> i. *Speed*
> ii. *Raising of C.O.G. by overloading*

3. What is meant by stable equilibrium as applied to rigid bodies.

> ***Solution:***
> *A rigid body is in stable equilibrium if after a slight displacement, the body returns to its original position.*

4. The figure below shows a burning candle standing on a bench.

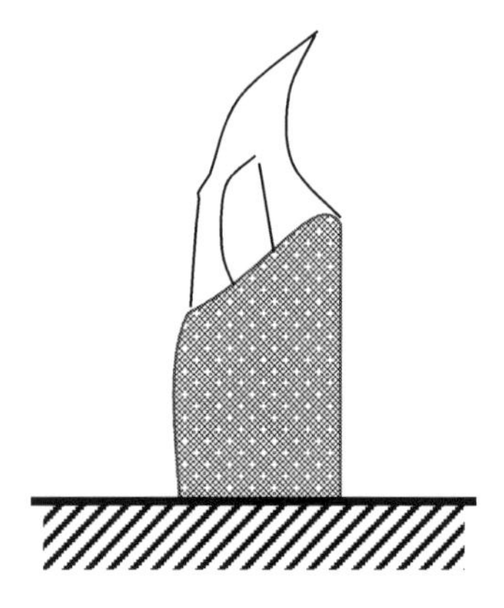

State and explain how the stability of the candle Standing on the bench changes.

> ***Solution***
> *As the candle burns, its stability increases (more stable). As the candle burns, it reduces in height hence reducing the centre of its gravity.*

5. Explain how the centre of gravity of a body affects its stability.

> ***Solution:***
> *The stability of a body increases with the lowering of he height of the centre of gravity.*

6. The figure shows a plastic bottle, which is weighted at A.

State and explain what happens when the bottle is displaced slightly and then released. (2 marks)

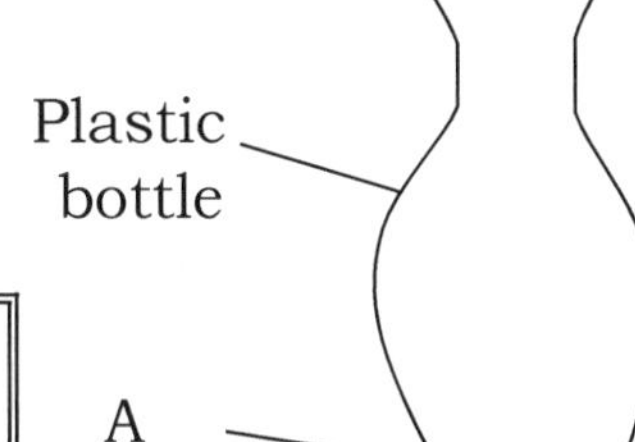

> ***Solution:***
> - *The bottle will swing sideways and then come to rest standing upright.*
> - *The weight at the bottom lowers its C.O.G. making it very stable.*

Chapter seven: Gas Laws

1. a. Explain how a rise in the temperature of gas causes a rise in the pressure of the gas if the volume is kept constant using Kinetic theory of gases.

   b. The figure shows a set up that may be used to verify Charles' law.

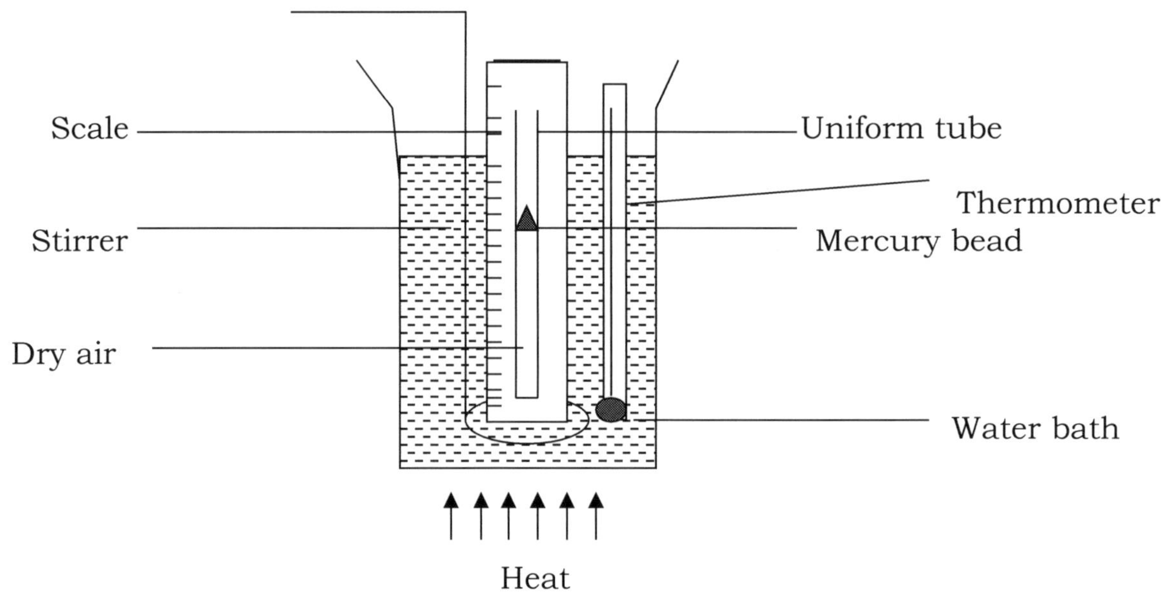

   i. State the measurements tat should be taken in the experiment.

   ii. Explain how the measurements taken in (i) above may be used to verify Charles law.

   c. A certain mass of hydrogen gas occupies a volume of $1.6m^3$ at a pressure of 1.5×10^5 Pa and temperature 12^0C. Determine its volume when the temperature is 0^0C at a pressure of 1.0×10^5Pa.

Solution:

a. A rise in temperature at constant volume increases the kinetic energy of
molecules hence rate of bombardment with the walls of container and force of bombardment. This is an increase in pressure.

b. i. Measurements to be taken are those of
- *Temperature*
- *Volume of gas in tube / height of the mercury bead.*

ii. *Measurement of volume (or length of dry air) in tube are taken with temperature at different intervals. Temperature change is achieved by heating. A graph of l (y-axis) against temperature in Kelvin is plotted and straight line drawn.*

c. $V_1 = 1.6m^3$, $P_1 = 1.5 \times 10^5 Pa$ and $T_1 = 12^0C = 285K$

$V_2 = ?$ $P_2 = 1.0 \times 10^5 pa$ and $T_2 = 0^0C = 273$

$$\frac{P_1V_1}{T_1} = \frac{P_2V_2}{T_2} \Rightarrow V_2 = \frac{P_1V_1T_2}{P_2T_1}$$

$$\therefore V_2 = \frac{1.5 \times 10^5 \times 1.6 \times 273}{1.0 \times 10^5 \times 285}$$

$$= \frac{655.2}{285}$$

$$= 2.299m^3$$

2. a. i. State pressure law of gases.

ii. A gas put in a container of fixed volume is subjected to a pressure of $4.2 \times 10^5 Nm^{-2}$ at a temperature of 37^0C. What will be the pressure of the gas at 80^0C?

b. In a pressure cooker, food is cooked rapidly. Explain why this is so.

Solution:

a. i. *The pressure of a fixed mass of a gas is directly proportional to its absolute temperature if its volume remains constant.*

ii.

$$\frac{P_1}{T_1} = \frac{P_2}{T_2}$$

$$\frac{4.2 \times 10^5}{310} = \frac{P_2}{353}$$

$$\therefore P_2 = 4.782 \times 10^5 N/m^2$$

b. *Pressure builds up hence raises the boiling point of water. Food cooks at a higher temperature.*

3. a. i. State Charles' Law on an ideal gas.

ii. A gas in a container at 27^0C has a volume of $50cm^3$ and a pressure of 750 mmHg. What will be the volume at s.t.p.?

iii. Show that pressure P of a fixed mass of a gas is directly proportional to density D.

b. The pressure, P, of a fixed mass of gas at constant temperature, T = 200K is varied continuously. The corresponding values of P and volume, V of the gas are as shown in the table below.

Pressure P(x 10^5 pa)	4.0	3.2	1.6	0.8	0.4
Volume V (x $10^{-9}m^3$)					

Given that $\frac{PV}{2R} = T$ where R is a constant, plot an appropriate graph and use it to determine the value of R.

Solution:

a. i. *The volume of a fixed mass of a gas at constant pressure is directly proportional to absolute temperature.*

ii. $\frac{P_1V_1}{T_1} = \frac{P_2V_2}{T_2}$

$\frac{750 \times 50}{300} = \frac{760V_2}{273} \Leftrightarrow V_2 = 44.9cm^3$

iii. $\frac{P_1V_1}{T_1} = \frac{P_2V_2}{T_2}$ and $V = \frac{m}{\rho}$ *where m is fixed mass of gas of density* ρ

$\Rightarrow \frac{P_1(m/\rho_1)}{T_1} = \frac{P_2(m/\rho_2)}{T_2}$

$\Leftrightarrow \frac{P_1}{T_1\rho_1} = \frac{P_2}{T_2\rho_2}$

$\therefore \frac{P}{T\rho} = constant$ i.e. $P \alpha T\rho$

4. a. State two reasons why the pressure of a gas increases when its temperature increases.

b. Differentiate between a real gas and an ideal gas.

c. The pressure of a gas is 4.0×10^5Pa when its volume is $2.7 \times 10^{-3}m^3$. Calculate the new pressure when the volume is reduced to $1.2 \times 10^{-3}m^3$, assuming that there is no change in the temperature.

Solution:

a. - *Rate of collision increases*
- *Average speed of particles increases*

b. *Ideal gas are those that maintain all properties of a gas at absolute zero temperature (-273^0C)*
Real gases to not exist at negative temperatures.

c. $P_1 = 4.0 \times 10^5 Pa$ $V_1 = 2.7 \times 10^{-3}m^3$ and $V^2 = 1.2 \times 10^{-3}m^3$.

$$P_1V_1 = P_2V_2 \Leftrightarrow P_2 = \frac{P_1V_1}{V_2} = \frac{4.0 \times 10^5 \times 2.7 \times 10^{-3}}{1.2 \times 10^{-3}}$$

$$P_2 = 9 \times 10^5 Pa$$

5. When cold at 15^0C, the pressure in a car tyre was measured at 200KPa. After a long run, the temperature of the tyre rose to 50^0C. If the volume of the tyre remained constant, find its pressure after the long run.

Solution:

$P_1 = 200KPa$ $P_2 = ?$

$T_1 = 15 + 273 = 288K$ $T_2 = 50 + 273 = 323K$

$$\frac{P_1}{T_1} = \frac{P_2}{T_2} \Leftrightarrow P_2 = \frac{P_1T_2}{T_1}$$ *[Pressure law]*

$$\therefore P_2 = \frac{200 \times 323}{288} = 224.3KPa$$

6. A small balloon is inflated and placed inside a bottle. Air is then pumped out of the bottle using a vacuum pump. State and explain what is observed as the bottle is being evacuated.

Solution:

- *The balloon increases in volume.*
- *The pressure outside the balloon is reduced while the pressure inside the balloon remains high, hence an increase in volume.*

7. The figure shows a graph of pressure P against volume V for a fixed mass of gas.

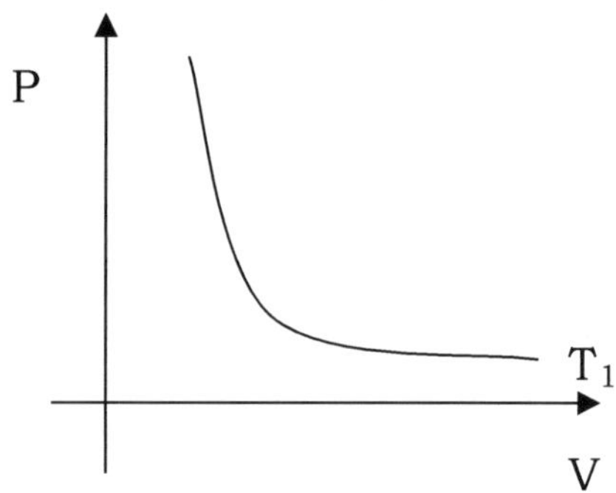

Sketch on the same diagram, the graph of pressure against volume for a temperature T_2, lower than T_1.

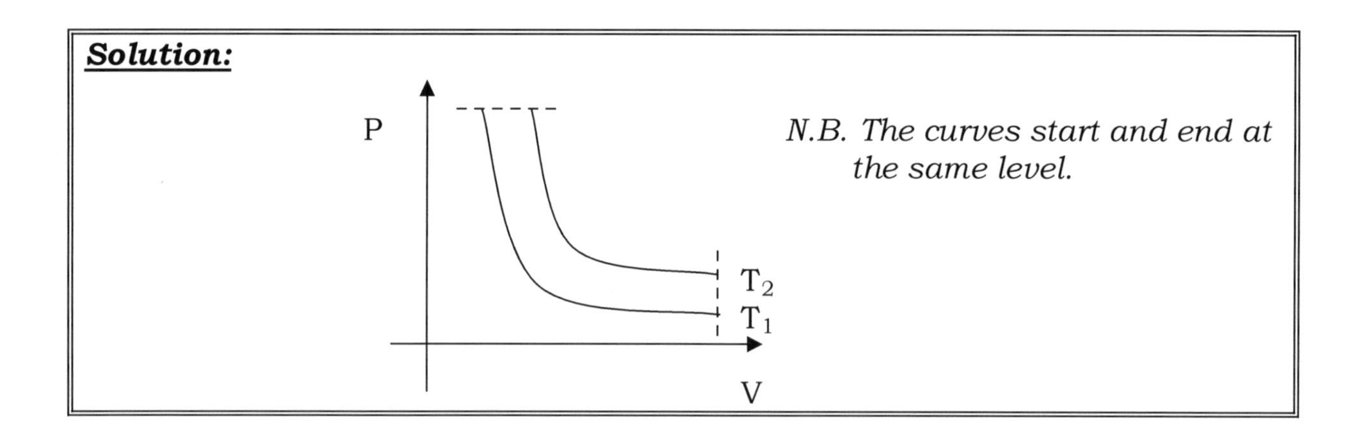

Chapter 8: Kinetic Theory of Matter

1. Pollen grains on water are observed to move randomly. Explain this motion.

Solution:

- *Invisible water molecules collide with the pollen grains making them to move randomly.*

2. An oil drop of volume $0.1mm^3$ spreads out to give a circular patch of area $800mm^2$. Estimate the size of a molecule of this oil.

Solution:

$$\text{Volume} = \text{Area} \times \text{Height}$$
$$0.1 = 800 \times h$$
$$\therefore \quad h = \frac{0.1}{800} = 1.25 \times 10^{-4}mm$$

3. Why is Brownian motion only exhibited by small particles?

Solution:
According to kinetic theory of matter, all particles have equal mean kinetic energies hence it is only the smaller particles that have noticeable speeds.

4. The figure shows drops of mercury, which were placed on a piece of glass by a student.

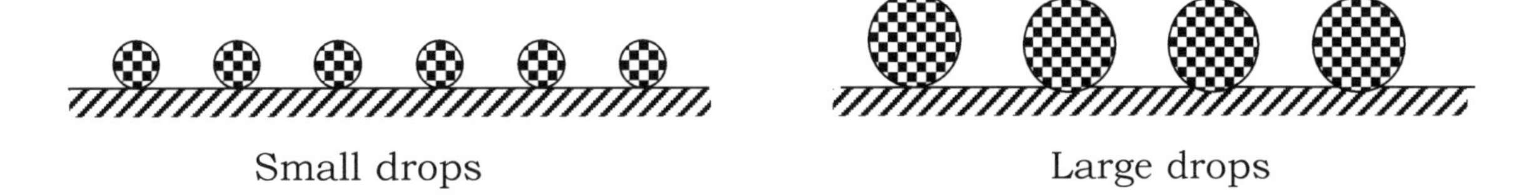

Small drops　　　Large drops

It was observed that the small drops were spherical while the large drops were cylindrical. Explain these observations.

Solution:
Cohesive force is overcome by weight of the bigger drops.

Chapter 9: Light

1. a. You are provided with two equilateral prisms and two convex lenses. Sketch a diagram showing how all the four can be arranged to make a simple prism binoculars.

b. State two conditions necessary for total internal reflection of light to occur.

c. The figure shows a human eye that has a defect.

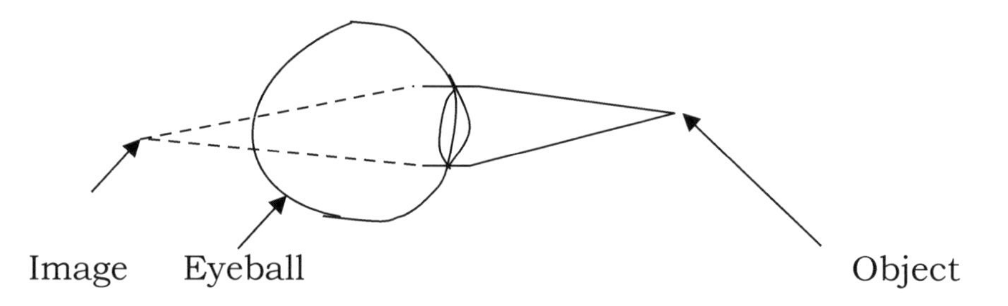

i. Name the defect.

ii. State one possible cause of the defect.

iii. Name the appropriate lens that can be used to minimize this defect.

d. Give one difference and one similarity between a lens of a camera and that of a human eye.

e. Differentiate between developing and printing of a photograph from a camera.

Solution:

a.

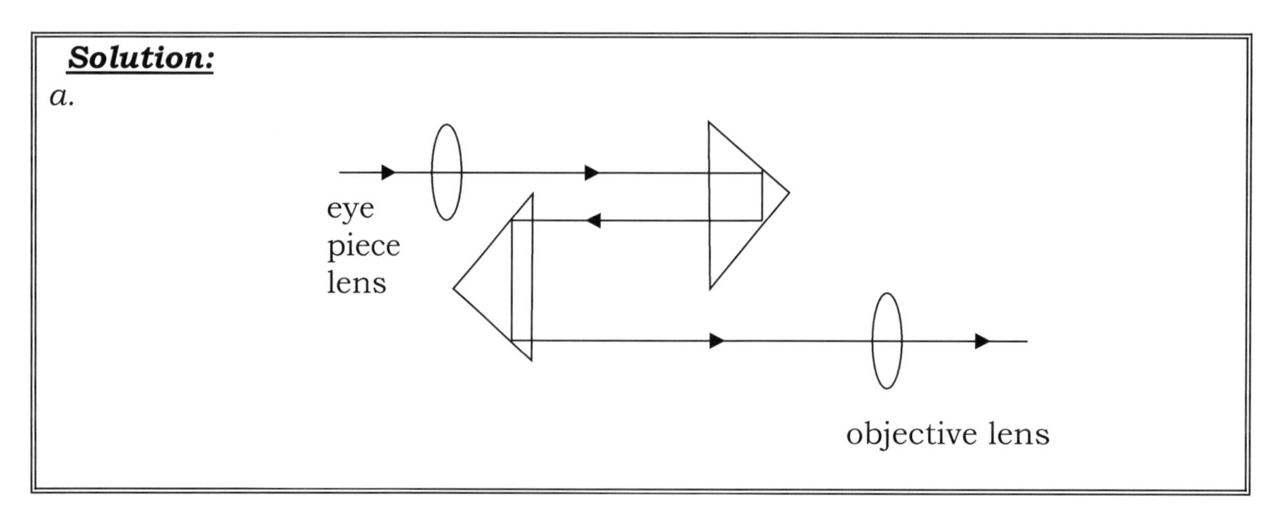

b. ⇔ *The angle of incidence should be greater than the critical angle of the first medium.*

⇔ *The first medium should be optically denser than the second medium.*

c. i. *Long sightedness*

ii. *The convex lens*

d. *Differences:*

Camera lens	Human eye
Has a shutter	No shutter
Glass lens	Crystalline lens
A constant focal length	Variable focal length

Similarities

- *Both have a light sensitive screen or vetting.*
- *Both have a mechanism for controlling amount of light entering them.*

e. *Developing involves the washing away of un-reacted chemical on a film to produce a negative while printing involves transferring the image from negative to paper as a photograph.*

2. A pin is placed at the bottom of a beaker containing a transparent liquid. When viewed from the top, the pin appears nearer the surface than it actually is.

a. Explain this observation.

b. Describe how the apparent depth of the liquid may be determined experimentally.

Solution:

a. *The observation is a result of refraction,*

- *The bending of light rays as they move from one medium to another.*
- *Change in velocity of light rays when they travel from one medium to another.*

b. *No parallax method: An optical pin is held along the outside wall of the beaker to appear to coincide with the one inside. Its position from the top of the liquid is measured which is the apparent depth.*

3. a. The figure below sows an arrangement of two convex lenses forming a compound microscope.

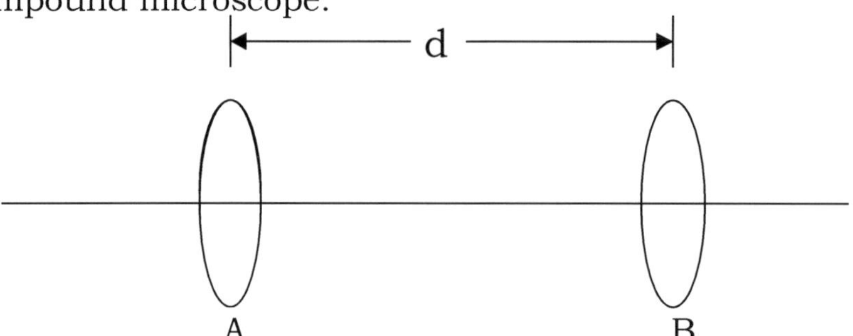

f_A = 10cm f_B = 20cm

An object is placed 15cm from lens A of focal length 10cm.

i. What is the position for the first image from lens A?

ii. If the distance of separation of the two lenses 'd' is 45cm. Determine the position of the final image due to the second lens B of focal length 20cm.

iii. What is the magnification of the final image?

b. i. State two conditions necessary for total internal reflection to occur.

ii. The figure shows an object O, placed in front of a concave mirror and its image, I formed by the mirror.

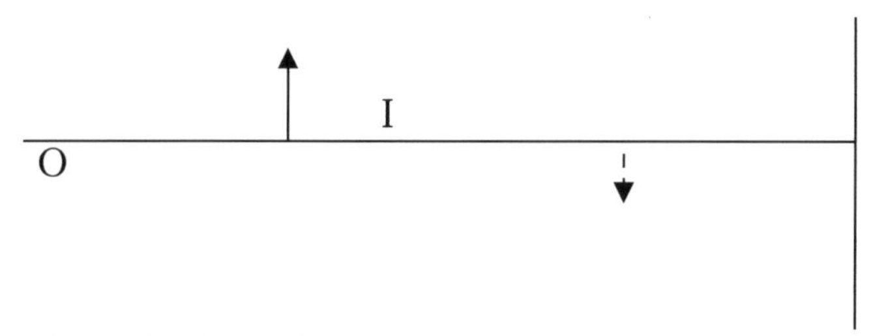

Sketch the rays to identify the principal focus 'f' of the concave mirror.

c. State the effect of the following on the image formed in a pinhole camera

i. Doubling the object distance.

ii. Doubling the size of the pinhole.

Solution:

a. i.

$$\frac{1}{f} = \frac{1}{u} + \frac{1}{v}$$

$$\frac{1}{10} = \frac{1}{15} + \frac{1}{v}$$

$$v = \frac{15 \times 20}{15 - 20} = 30cm$$

ii. $u_B = 15cm$ and $f = 20cm$

$$\frac{1}{20} = \frac{1}{15} + \frac{1}{v}$$

$$v = \frac{15 \times 20}{15 - 20} = -60cm$$

iii. $M = M1 \times M2$

$$= \frac{30}{15} \times \frac{60}{15}$$

$M = 8$

b. i. *Conditions for total internal reflection to occur*

- *Light must be travelling from a more optically dense medium to a less dense medium.*
- *The angle of incidence must be greater than the critical angle of the dense medium.*

ii.

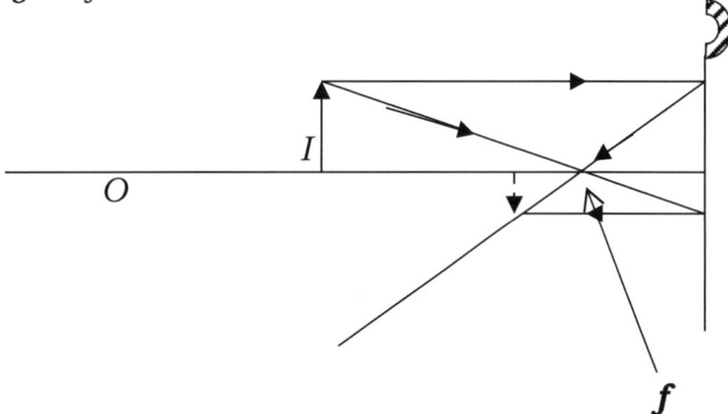

c. i. *Image height is halved.*

ii. *Image is blurred*

4. a. O is an object and PQ is a plane mirror.

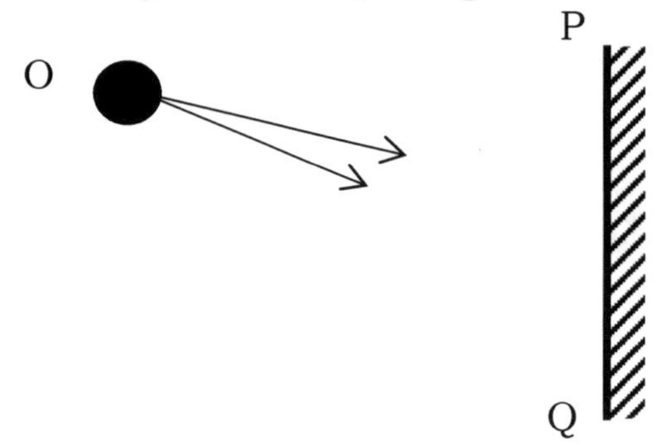

Complete the path of the two rays from object O and mark the final position of the image I.

b. You are provided with the following apparatus:

- Retort stand with a clamp
- Optical pin
- Convex lens
- Plane mirror
- Metre rule

With an aid of a diagram, describe how you can determine the focal length of a convex lens.

c. Calculate the magnification of a real image formed 25cm from converging lens of focal length 10cm.

Solution:

a.

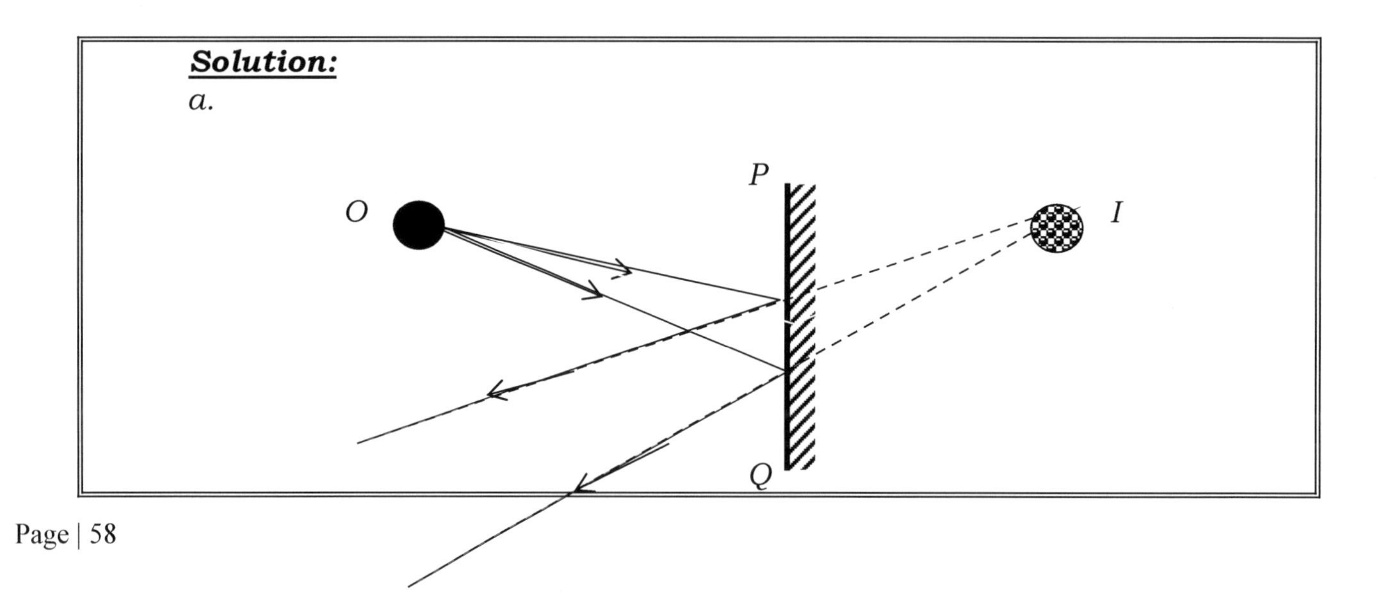

b. *Determination of focal length of a convex lens:*

- *Set up the apparatus as shown in the diagram.*
- *Adjust the optical pin by moving it to and fro until its image coincides with the object pin by no parallax method.*
- *Measure the distance between the lens and the pin.*
- *Repeat several times.*
- *Find the average distance.*
- *The average distance obtained is the focal length of the lens.*

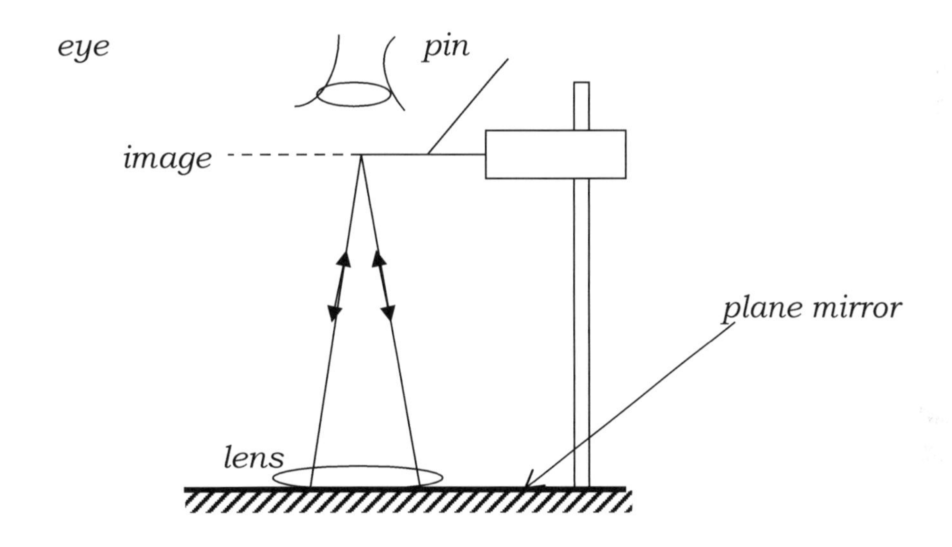

c. $M = \frac{v}{f} - 1 = \frac{25}{10} - 1 = 1.5$

or

$\frac{1}{u} + \frac{1}{v} = \frac{1}{f}$ $\qquad 1 + \frac{v}{u} = \frac{v}{f}$

$M = \frac{v}{f} - 1 = \frac{25}{10} - 1 = 1.5$

5. a. i. Distinguish between lunar and solar eclipses.

ii. The figure shows an incident ray at an angle of 30^0 to the mirror.

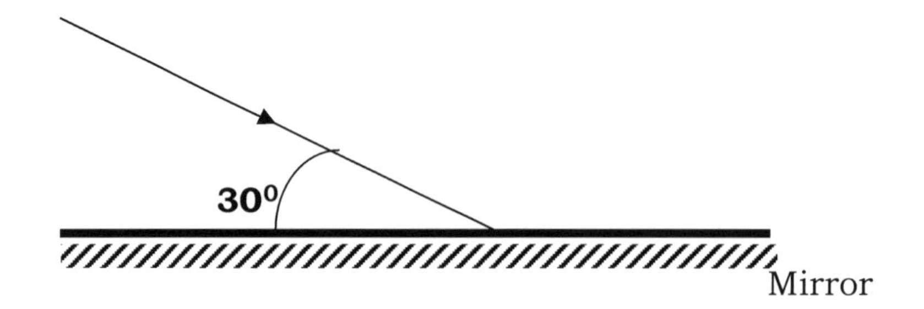

Find the angle of reflection if the mirror is rotated at an angle of 10^0 clockwise.

b. An object is placed 10cm in front of a concave mirror of radius of curvature of 12cm. By using a ray diagram,

i. Determine the position of the image.

ii. State the nature of the image formed.

Solution:

a. i. Lunar - Eclipse of the moon.
Earth between the sun and the moon.

Solar - Eclipse of the sun
Moon between the sun and the earth.

ii. New angle after rotation $= 30^0 - 10^0 = 20^0$

$\therefore$ *New angle of incidence* $= 90^0 - 20^0 = 70^0$

In reflection, angle of incidence = angle of reflection
$\therefore$ *New angle of reflection =* 70^0.

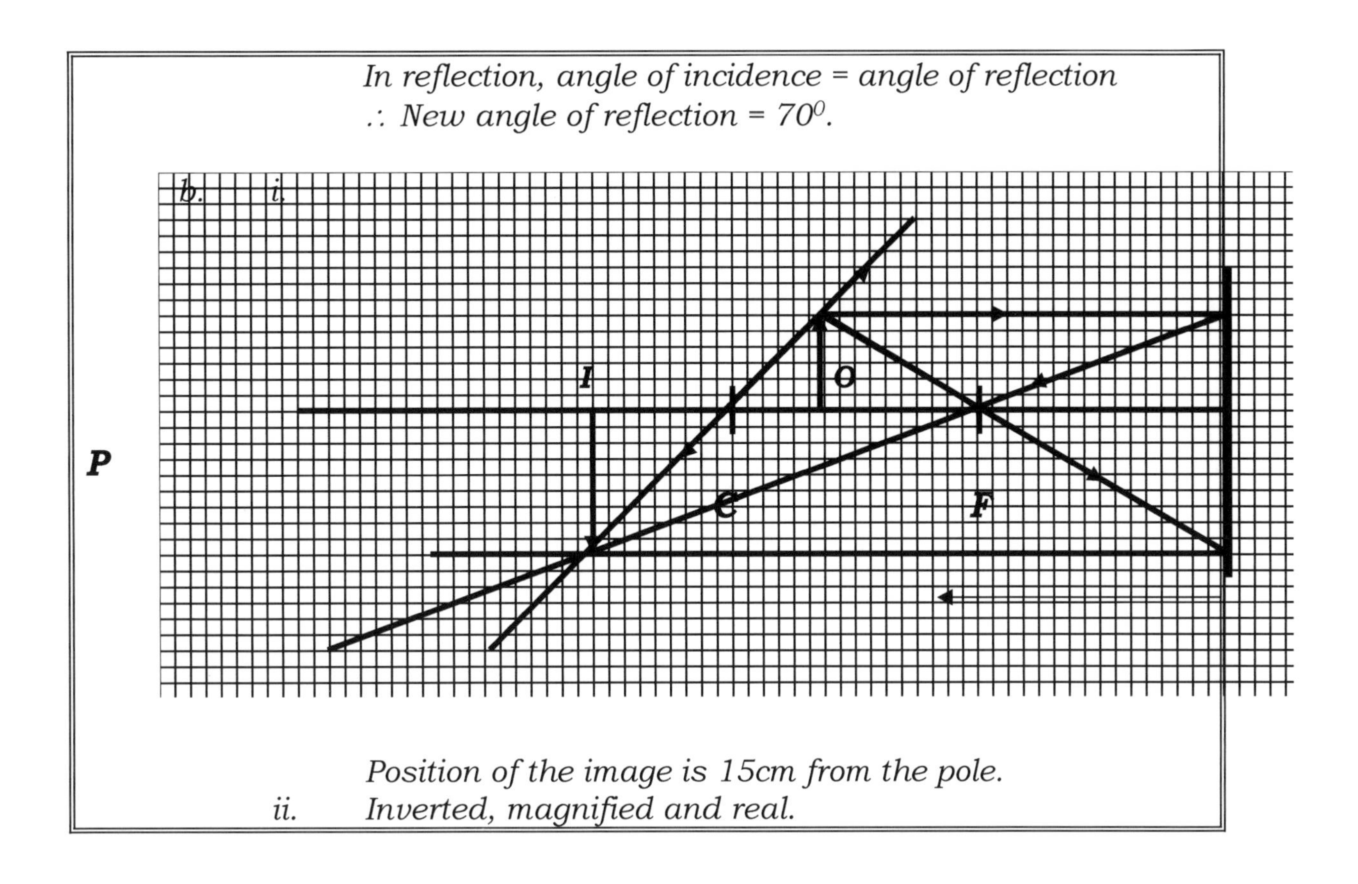

Position of the image is 15cm from the pole.
ii. *Inverted, magnified and real.*

6. a. i. State Snell's law.

ii. The figure shows a 6cm layer of water (n = 1.33) over a 2cm
layer
of glycerine (n = 1.47).

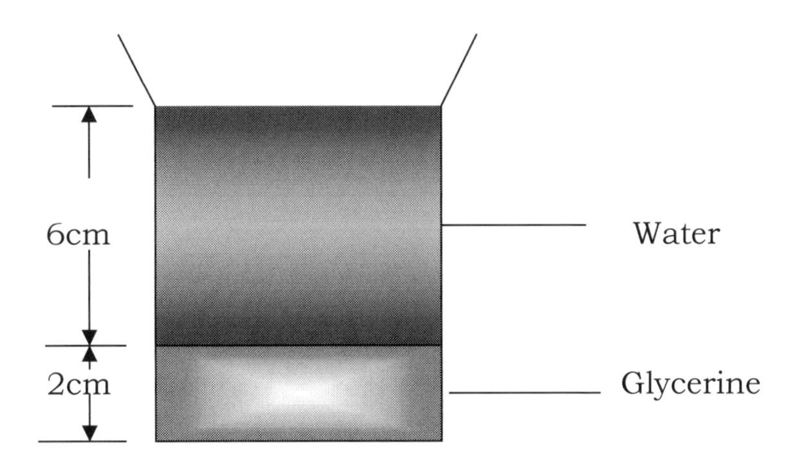

What is the apparent distance from the top of the surface of water to the bottom of the glycerine layer when viewed vertically downwards on normal incidence?

b. The figure shows an object placed in front of the objective lens of a microscope. The focal length of the objective lens is 2.ocm while the focal length of the eyepiece is 5.0cm. The two lenses are 15cm apart and the final image is formed 25cm from the eyepiece lens. (Figure not drawn to scale).

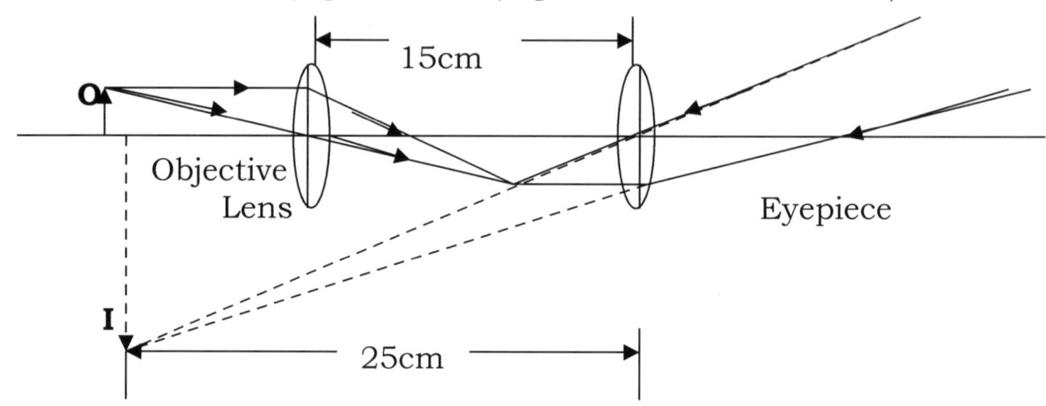

Determine the position of the object from the objective lens.

c. The figure shows a virtual image, I, formed by a convex mirror.

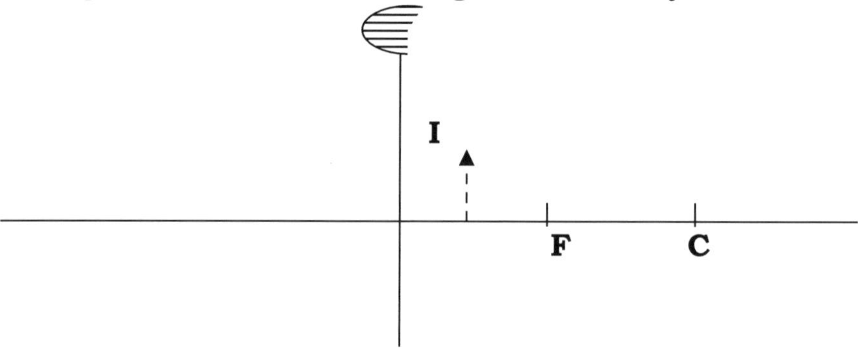

On the diagram, draw appropriate rays to locate the position of the object.

Solution:

a. i. *For a ray of light passing from one medium to another, the sine of the angle of incidence is proportional to the sine of the angle of refraction.*

ii. *Apparent depth in glycerine = 2.0/1.47 ≃ 1.36cm*
Apparent depth in water = 6/1.33 ≃ 4.51cm
Total apparent depth = 5.87cm.

b. *In the eyepiece, the image of the objective lens is its object.*

$\Rightarrow \quad \frac{1}{-25} + \frac{1}{u} = \frac{1}{5} \quad \Leftrightarrow \quad \frac{1}{u} = \frac{1}{5} + \frac{1}{25}$

$\Rightarrow \quad \frac{1}{u} = \frac{6}{25} \quad \therefore \quad u = 4^1/_6$*cm from the eyepiece.*

In the objective, image distance = $10^5/_6$

$\therefore \quad \frac{6}{65} + \frac{1}{u} = \frac{1}{2}$

$\Leftrightarrow$ *u = 2.5cm infront of the objective lens.*

c.

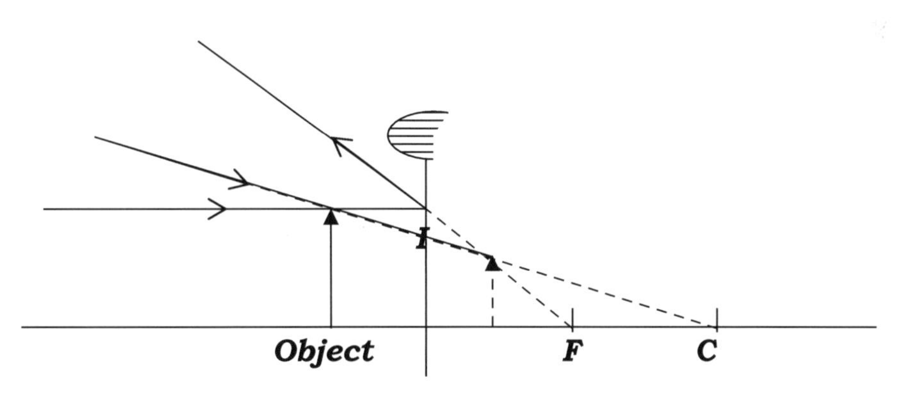

7. The figure shows a ray of light incident on water air boundary from the side of water.

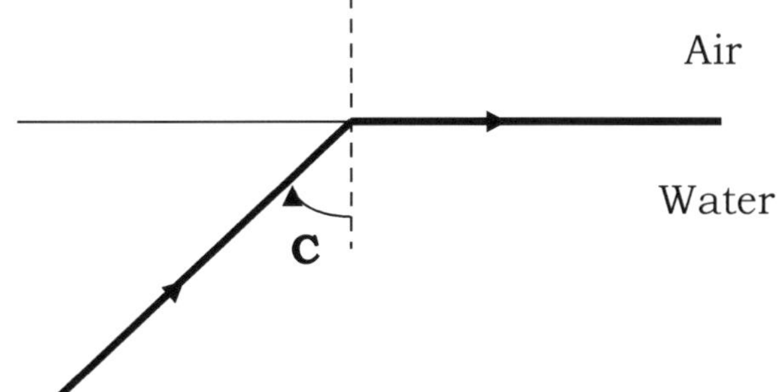

a. If the refractive index of water is 4/3, find the critical angle.

b. Draw sketches to show what happens when the light is now at 20^0 and 60^0. Account for the differences in the two cases.

c. In an experiment to measure real and apparent depth of water, the following results were obtained.

Real Depth (cm)	8.1	12.0	16.0	20.0
Apparent Depth (cm)	5.9	9.1	12.0	15.1

i. Plot a graph of real depth (y-axis) against apparent depth.

ii. Determine the refractive index from the graph.

Solution:

a. *Refractive index* $= \dfrac{1}{\sin C}$ *where C is the critical angle.*

$\Leftrightarrow \quad \sin C = \dfrac{1}{4/3} = 0.75$

$\therefore \quad C = 48.59^0$

b. *At* 20^0

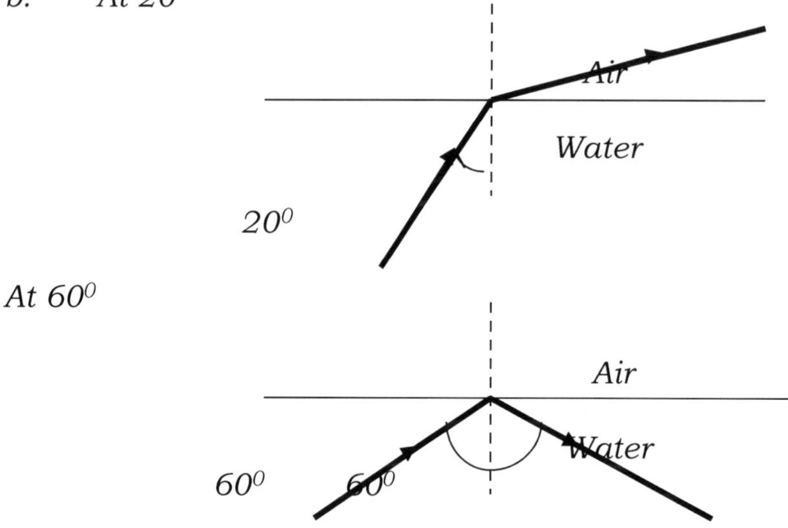

At 20^0, there is refraction since the critical angle, C = 48.59^0 is not exceeded but incident at 60^0, there is total internal reflection in water since the critical angle is exceeded.

c. i.

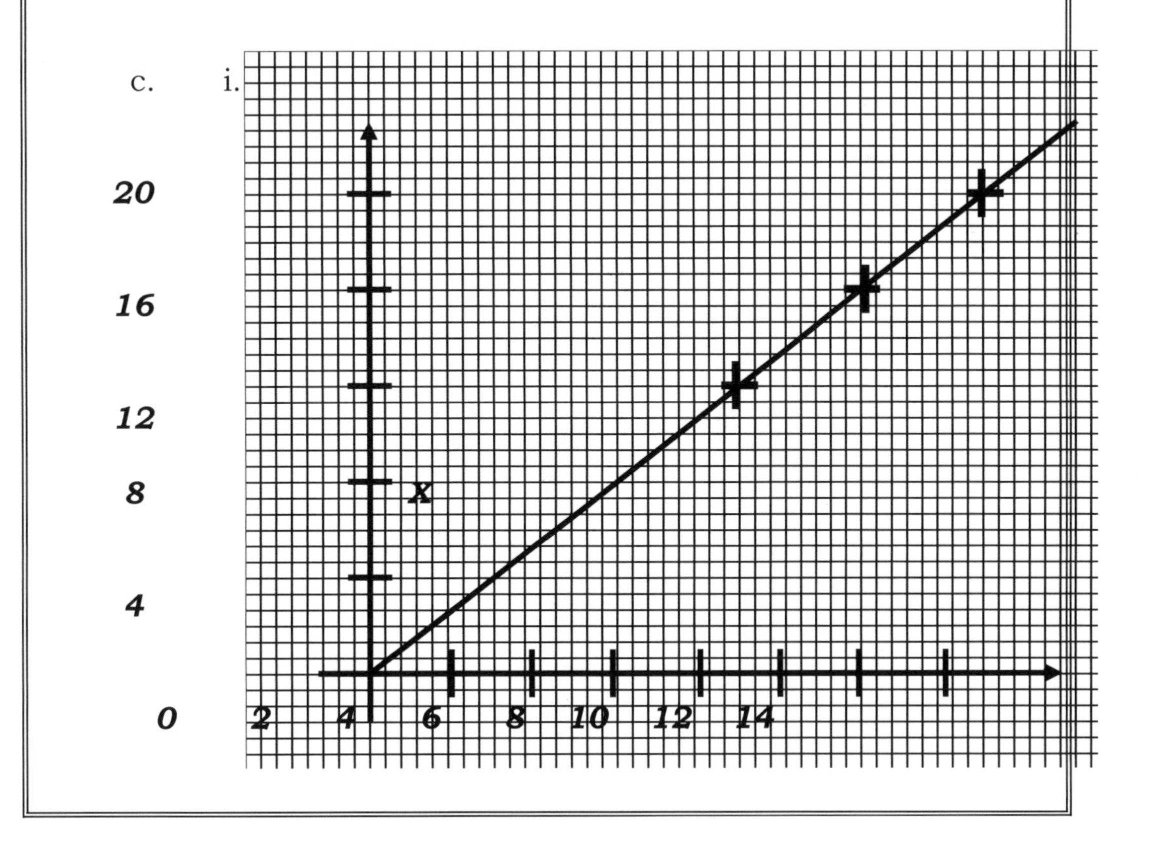

ii. *Refractive index = gradient of graph*

$\Leftrightarrow$ *Gradient* = $\dfrac{20 - 8.1}{15.1 - 5.9} \simeq 1.3$

$\therefore$ *Refractive index = 1.3*

8. a. i. State two conditions necessary for refractive of light to occur.

ii. Draw a pair of rays of light to illustrate how the ring at the bottom of the pond is seen by the observer in the figure.

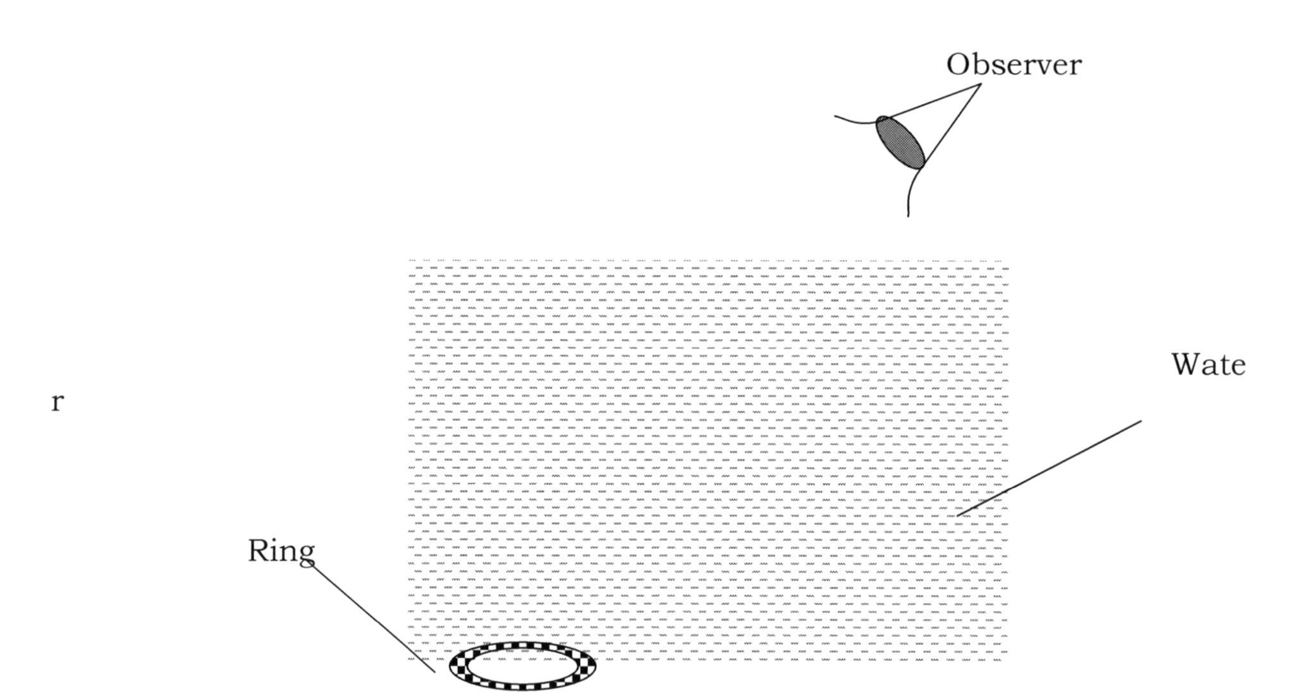

b. In an experiment to determine the refractive index of a liquid, the displacement of a coin at the bottom of various heights of the liquid were found to be as in the table below.

Height, H of liquid (cm)	40	34	30	22	15	10
Displacement, d (cm)	8	6.8	6	4.4	3	2

i. Plot a suitable graph and use it to determine the refractive index of the liquid.

ii. Draw a suitable diagram for the arrangement of the apparatus for the above experiment.

Solution:

a. i. - *Ray of light must move across the boundary between two optical media of different optical densities.*
- *A ray should strike the boundary at an angle.*

ii.

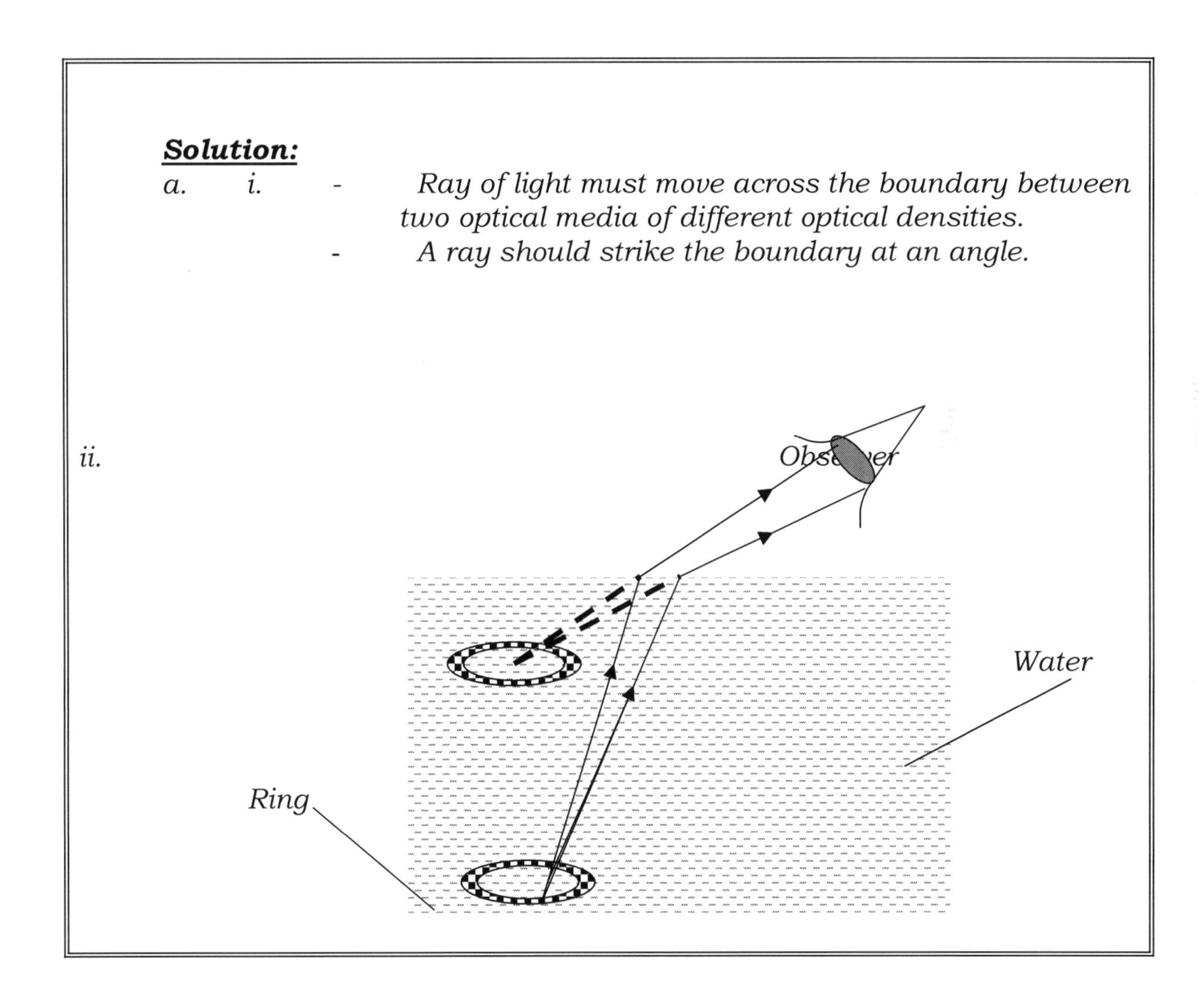

b. i. Refractive index $n = \frac{\text{Real depth}}{\text{Apparent depth}}$

∴ Slope of a graph of real depth vs apparent depth = n

Height, H of liquid (cm)	40	34	30	22	15	10
Displacement, d (cm)	8	6.8	6	4.4	3	2
Apparent depth (cm)	32	27	24	18	12	8

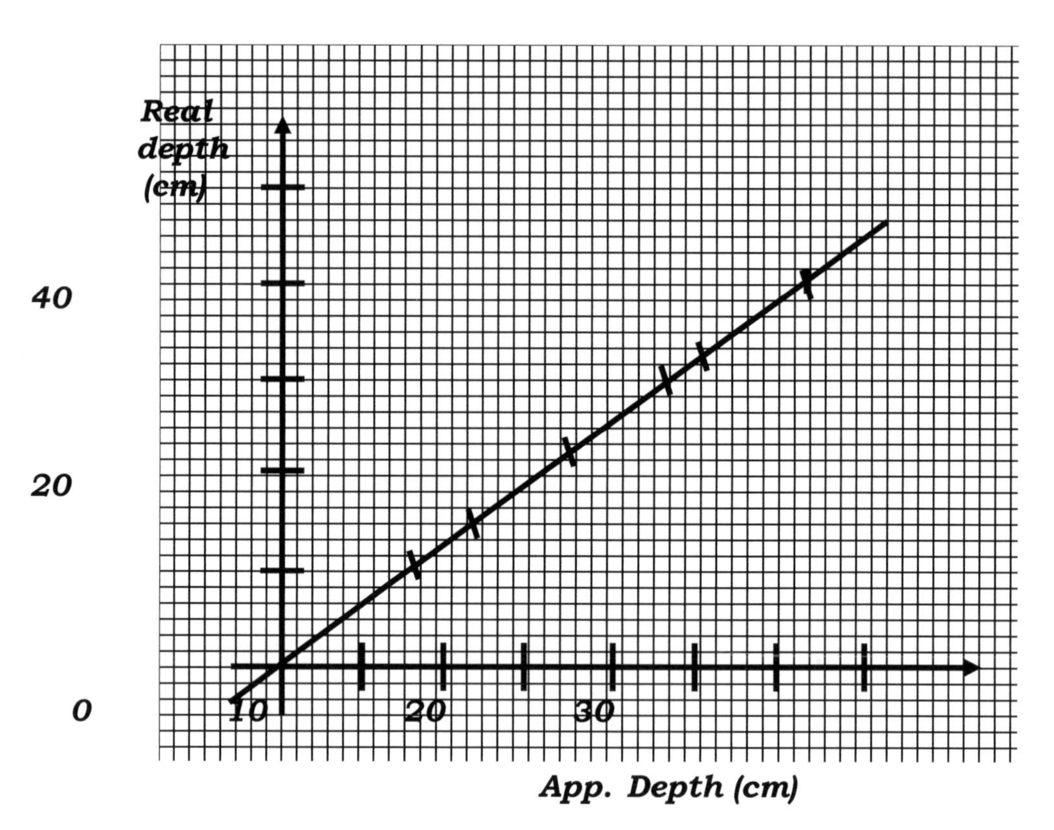

ii.

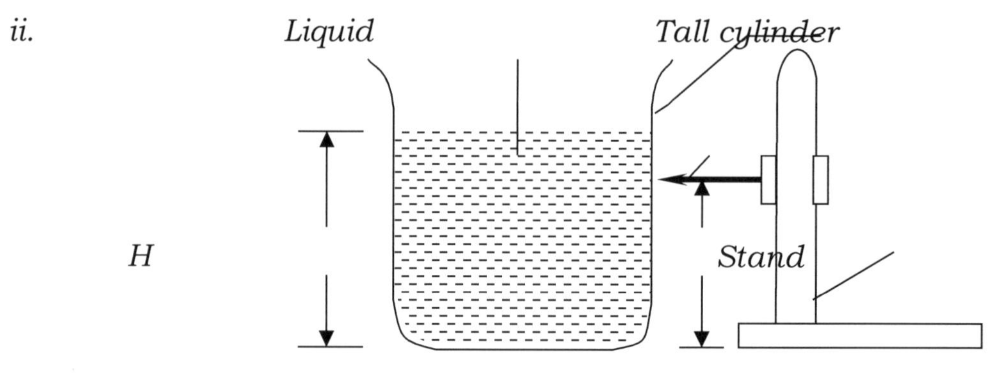

9. Light travels through liquid of refractive index of $^4/_3$. Determine the velocity of light in the liquid given that the velocity in air = 3.0×10^8m/s.

Solution:

Refractive index = $\frac{\text{Velocity of light in air}}{\text{Velocity of light in liquid}}$

$\Rightarrow \quad \frac{4}{3} = \frac{3.0 \times 10^8}{V_L}$

$\Leftrightarrow \quad V_L = 3.0 \times 10^8 \times \frac{3}{4} = 2.25 \times 10^8 m/s$

10. a. The table shows values of magnification, m and the image distance, V for a convex lens.

Magnification, m	0.09	0.4	0.78	1.49	1.74
Image distance, V (cm)	4.41	5.62	7.1	10.05	11.10

i. Plot a graph of m (y-axis) against V.

ii. Given the equation $m = Vf^{-1}$, determine the focal length, f from the graph.

b. i. A concave lens forms an image 30.0cm away when the object is at a distance of 7.5cm. What is its focal length?

ii. Name the eye defect that a concave lens can be used to correct. Explain.

11. In a class experiment on thin lenses, the following data was collected.

Object distance, U (cm)	30	40	60	75	90
Image distance, V (cm)	30	24	20	18.75	18
VU (cm^2)					
V + U (cm)					

i. Complete the table.

ii. Plot a graph of VU (y-axis) against V + U.

iii. Determine the gradient of the graph.

iv. Given that the focal length of the lens is given by

$$F = \frac{UV}{U + V}$$

Find the power of the lens.

c. The table below shows the results obtained when such an experiment was performed using various depths of a liquid.

Apparent depth (cm)	2.21	3.68	5.15	6.62	8.09	9.50
Real depth (cm)	3.0	5.0	7.0	9.0	11.0	13.0

i. Plot a graph of the apparent depth (y-axis) against the real depth.

ii. Using the graph, determine the refractive index of the liquid.

12. A girl observes her face in a concave mirror of focal length 90cm. If the mirror is 70cm away from her face, state two characteristics of the image observed.

> ***Solution:***
> - *Vitual*
> - *Magnified* } *N.B: Object is between optical centre and principle focus.*
> - *Upright*

13. The diagram shows the image I of an object placed in front of a mirror PQ.

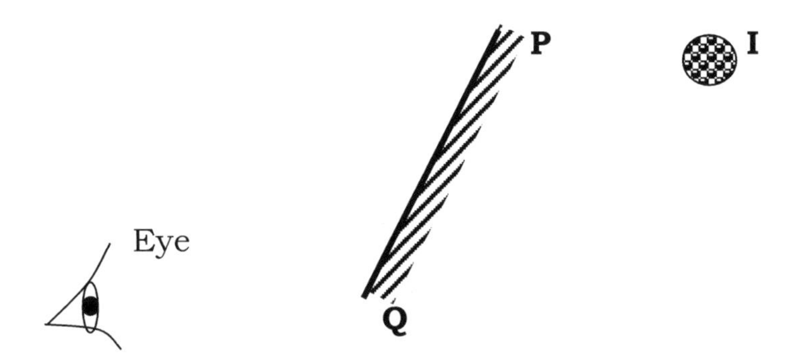

By ray diagram, locate the position of the object.

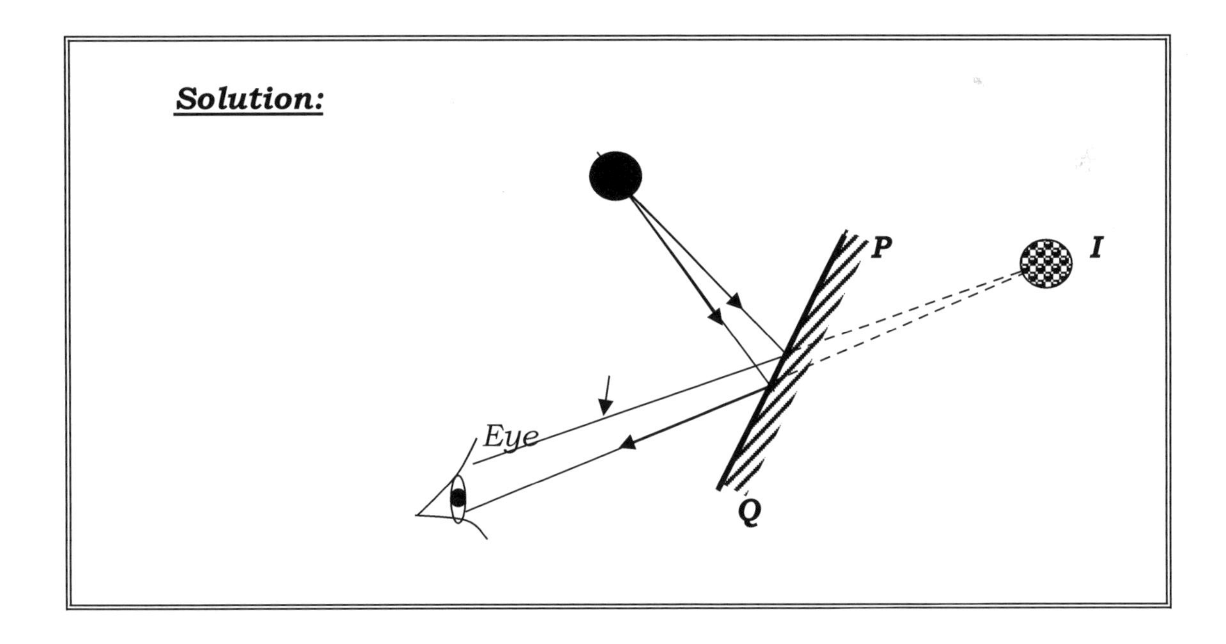

14. Use a ray diagram to show how a converging mirror forms a virtual image.

Solution:

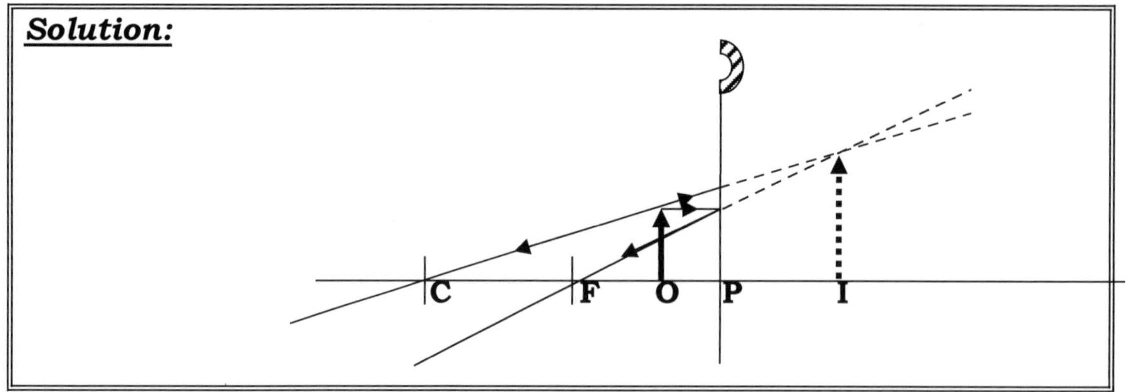

15. The figure shows a point object on a wall being focused using a concave mirror. P is its pole.

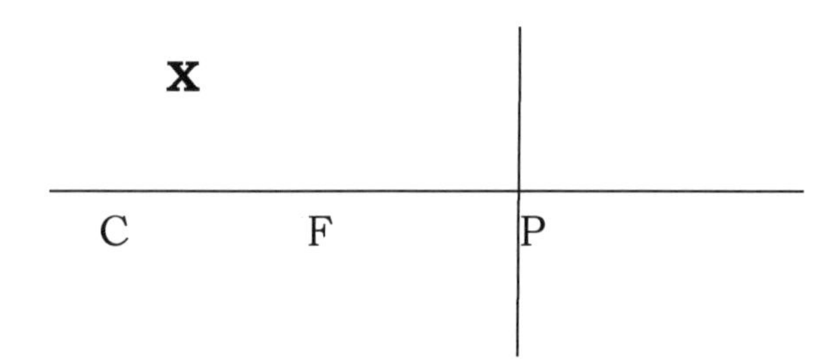

Draw appropriate rays to locate the image of the object.

Solution:

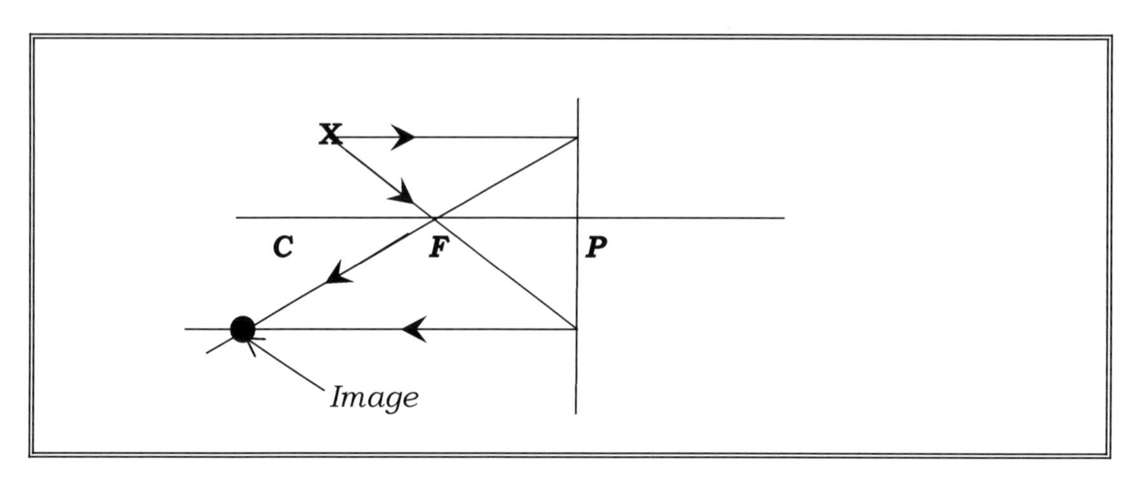

16. The refractive index of medium 1, n_1 = 2.42 and the refractive index of medium 2, n_2 = 1.5. Calculate the refractive index for a ray moving from medium 1 to medium 2.

Solution:

$${}_1n_2 = \frac{1}{n_1} \times n_2$$

$$= \frac{1.5}{2.42} = 0.6198$$

Chapter 10: Linear and Circular Motion, and Newton's Laws of Motion

1. a. Define force in terms of momentum.

   b. A car of mass 1,000kg travelling at 20m/s on a horizontal road is brought to rest in a distance of 40m by the action of brakes and friction force. Find

      i. The deceleration

      ii. The average stopping force.

      iii. The time taken to stop the car.

   c. The figure below shows a block of mass 20kg pulled along a horizontal surface by a force of 150N inclined at 60^0 to the horizontal. The block moves 3m horizontally.

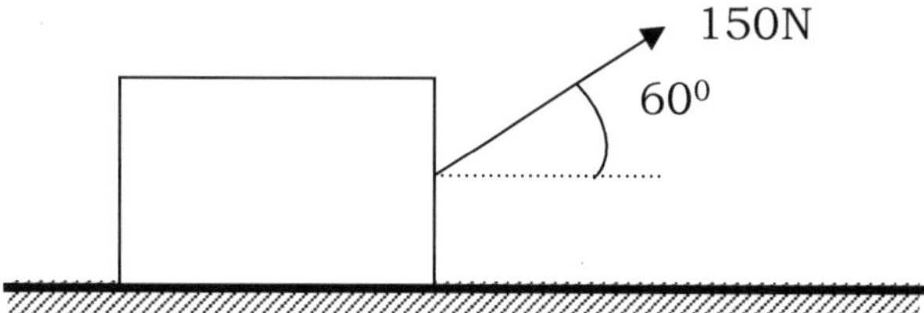

Given that the coefficient of kinetic friction between the surface and the mass is 0.2, determine

   i. The horizontal component of the pulling

   ii. The resultant force acting on the mass horizontally.

   iii. The useful work done on the block.

Solution:

a. *Force is the rate of change of momentum.*

b. i. $v^2 = u^2 + 2as$

$0 = 400 + 2 \times a \times 40$

$a = -5ms^{-2}$

i.e. a deceleration $= 5ms^{-2}$

ii. *Stopping force,* $f = ma$

$\Rightarrow \quad f = 1{,}000 \times -5$

$f = -5{,}000N$

iii. $v = u + at$

$0 = 20 - 5t$

$t = 4s$

c. i. *Horizontal component* $= 150\cos 60^0 = 75N$

ii. *Friction force* $= 0.2 \times 200 = 40N$

Resultant force $= 75N - 40N = 35N$

iii. *Useful work done* $= 35 \times 3 = 105J$

2. The figure below shows a system of forces on a bench.

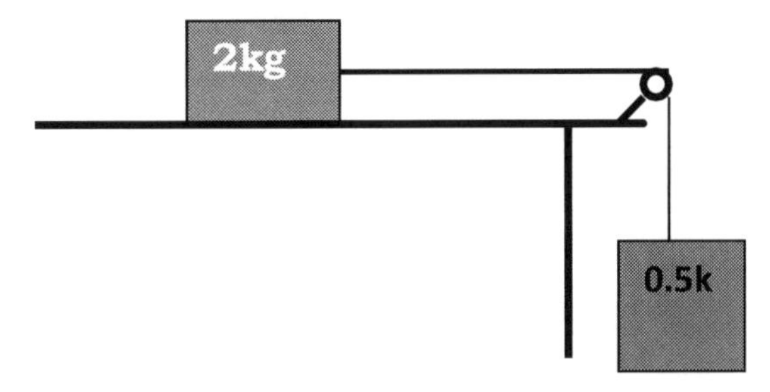

The frictional force between the bench and block of 2kg is 3N. The block when released, move through a distance of 0.6m.

Calculate

i. Work done against friction.

ii. Tension on the string

Solution:

i. *Work done = force x distance*
$= 3 \times 0.6 = 1.8J$

ii. *Tension* $T = ma - F_r$ *where* F_r *is frictional force.*
$= 2 \times 8 \times {}^{2}/_{3} - 3$
$= 7.6N$

3. a. i. State the principle of conservation of linear momentum.

ii. Two bodies of masses 2,900kg and 2,000kg are travelling in opposite directions with velocities 20m/s and 15m/s respectively. They collide head on and stick to each other. Determine the speed of the bodies after collision.

b. A potter's wheel of diameter 0.3m spins horizontally about a vertical axis as shown in the figure.

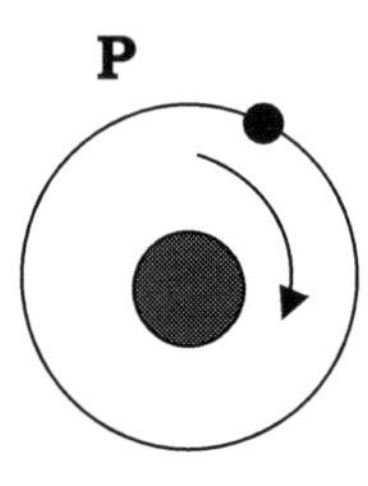

P is a particle of clay stuck to the edge. If the wheel rotates at 240 revolutions per minute, calculate

i. The angular velocity of the wheel.

ii. The acceleration of the clay particle P.

Solution:

a. i. If no external forces act on a system of colliding objects, the total momentum of the objects in a given direction remains constant.

ii. $m_1u_1 + m_2u_2 = (m_1 + m_2)v$

$2,900 \times 20 + 2,000 \times -15 = (2,900 + 2,000)v$

$4,900v = 58,000 - 30,000$

$v = 5.71 m/s$

b. i. $\omega = 2\pi/T = 2\pi f$

$\therefore \omega = 2 \times \pi \times \frac{240}{60}$

$= 25.12 rad/sec.$

ii. $a = \omega^2 r$

$= (25.12)^2 \times 0.15$

$= 94 m/s^2$

4. a. All passengers in public service are required to wear safety belts. Explain using Newton's law of motion how a safety belt can reduce injuries in case of an accident.

b. A trolley of mass 3kg moving at 6m/s to the right collides with another trolley of mass 2kg moving at 3m/s in the opposite direction. If the collision is perfectly elastic find their velocity after collision and state their direction.

Solution:

a. In case of an accident/sudden stop, the passengers would want to continue moving (Inertia). The belts slow down/stops the forward motion.

b. Momentum before collision = momentum after collision

Taking the motion to the right to be +ve,

$(3 \times 6) + (2 \times -3) = (3 + 2)v$

$\Rightarrow \quad 5v = 12$

$\Leftrightarrow \quad v = 2.4 m/s$

Since the velocity is +ve, they will move to the right.

5. a. i. Define acceleration and state its SI units.

ii. A train slows from 108km/h with a uniform deceleration of $2 m/s^2$. How long will it take to reach 18km/h?

b. i. State the Bernoulli's principle.

ii Water at a speed of 2m/s is pumped through a horse pipe of diameter 2.8cm to a sprinkler having 21 holes, each with a diameter of 1.4mm. What is the speed of delivery of the water from the sprinkler?

Solution:

a. i. Acceleration is the rate of change of velocity.
Its SI unit is metre per second squared (m/s^2)

ii. $u = 108km/h = 30m/s$
$v = 18km/h = 5m/s$
$a = -2m/s^2$

$$a = \frac{v-u}{t} \Leftrightarrow t = \frac{v-u}{a} = \frac{5-30}{-2}$$

$t = 12.5sec.$

b. i. Provided a fluid is non viscous, incompressible and its flow streamline, an increase in velocity produces a corresponding decrease in the pressure it exerts.

ii. $A_1V_1 = A_2V_2 \Rightarrow V_2 = \frac{A_1V_1}{A_2}$

$$V_2 = \frac{\Pi \times (0.014)^2 \times 2}{\Pi \times 21 \times (0.0007)^2} = 38.09m/s$$

6. a. What is a centrifuge?

b. Convert 3.6 x 10^3 revolutions per hour into radians per seconds.

c. A pendulum bob of mass 1kg is attached at one end of string. The bob moves in horizontal circle in such a way that the string is inclined at 30^0 to the vertical. Calculate the tension in the string.

Solution:

a. *An apparatus that separates a mixture of substances of different densities.*

b. $3.6 \times 103\ rev/h = \frac{3.6 \times 10^3}{60 \times 60}\ rev/s = 1 rev/s$

but $1\ rev/s = 360^0/s = 2\Pi rad/s = 2 \times 3.14 = 6.28 rad/s$

c.

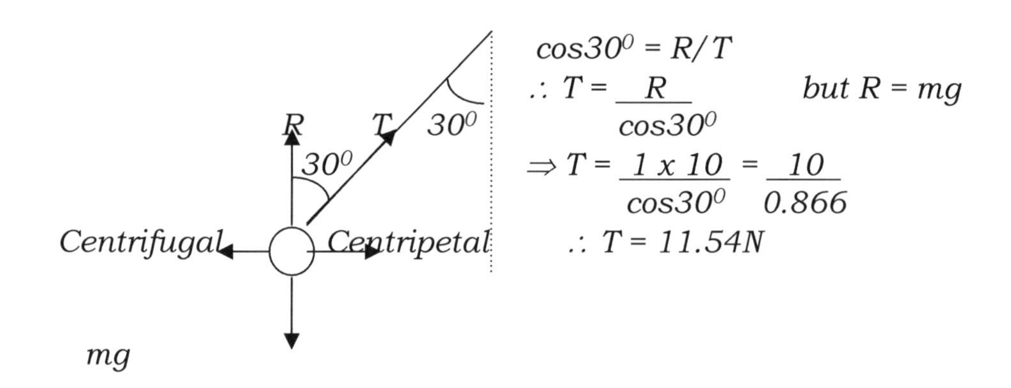

$\cos 30^0 = R/T$

$\therefore T = \frac{R}{\cos 30^0}$ but $R = mg$

$\Rightarrow T = \frac{1 \times 10}{\cos 30^0} = \frac{10}{0.866}$

$\therefore T = 11.54N$

7. a. Define centripetal force.

b. The table below shows the results of an experiment to establish the relationship between the centripetal force, F, and the radius, r, of the circle in which the body of mass, m rotates.

Number of 100g masses	8	7	6	5	4	3
Radius, r (cm)	70	61	52	43	34	25
Force F due to masses						

i. Complete the table

ii. On a grid, plot a graph of F (y-axis) against radius r, in meters.

iii. Using your graph, determine the angular velocity, ω, when the mass m is 150g.

c. A turntable is rotating at sixty revolutions per minute.

i. What is its angular velocity?

ii. Given that the radius of the turntable is 0.2m, what is the liner velocity of a particle on the edge of the turntable?

8. a. Two blocks of masses 20kg and 30kg are suspended at the ends of a weightless inextensible string passing through a frictionless single fixed pulley as shown in the diagram.

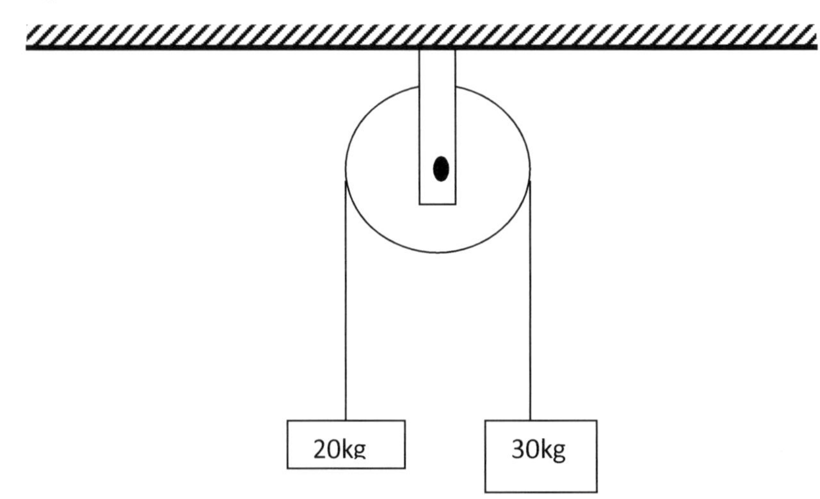

i. If both are set free, determine the acceleration of the system.

ii. If the 30kg is rested on a platform so that the system is at rest and the 20kg is replaced by a monkey of the same mass, at what acceleration must the monkey climb up the string to just lift the block off the platform?

b. The velocity-time graph of the motion of particle under gravity is as shown below.

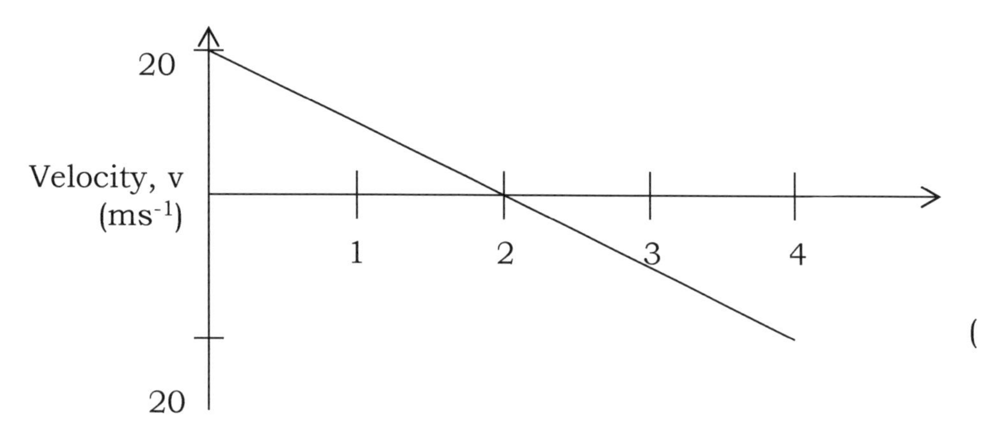

Time, t

s)

(

i. Briefly, describe the motion of this particle.

ii. Evaluate the total area under this graph.

iii. What is the physical meaning of your results in b (ii)?

1v. Sketch a displacement-time graph of the particle in b (i).

c. The diagram below shows an experiment set up by a form four student

to investigate the relationship between centripetal force and radius in

uniform circular motion.

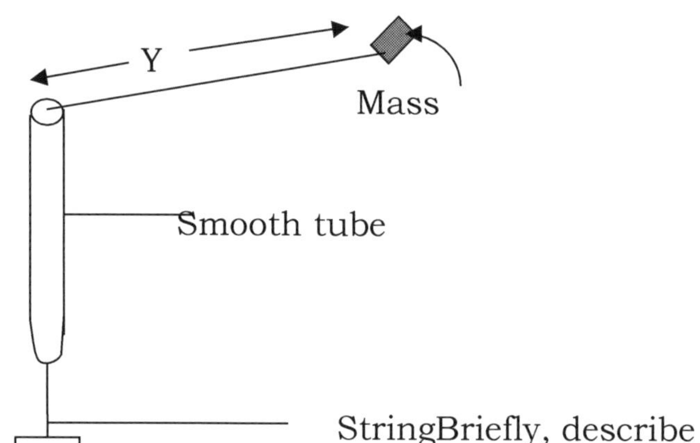

Briefly, describe how this set up can be used to obtain suitable values for this investigation.

Solution:

a. i. Taking that the system accelerates at a m/s^2 clockwise

and the tension on the string is T, then

from F =ma $T - 200 = 20a$

and $300 - T = 30a$

By adding the two equations,

$100 = 50a \quad \therefore a = 2m/s^2$

ii. The acceleration of the monkey must produce a force of

$(300 - 200) = 100N$ to balance the block.

From F = ma, $100 = 20a \Leftrightarrow a = 2m/s^2$

b. i. The particle is thrown upwards with a velocity of 20m/s

and decelerates to a momentary stoop after 2 seconds

before accelerating back to the ground having acquired a

velocity of -20m/s on impact.

ii. Total area = $(\frac{1}{2} \times 20 \times 2) + (\frac{1}{2} \times (-20) \times 2)$

$= 20 - 20 = 0.$

iii. It means that after 4 seconds, the displacement of the particle is zero.

iv.

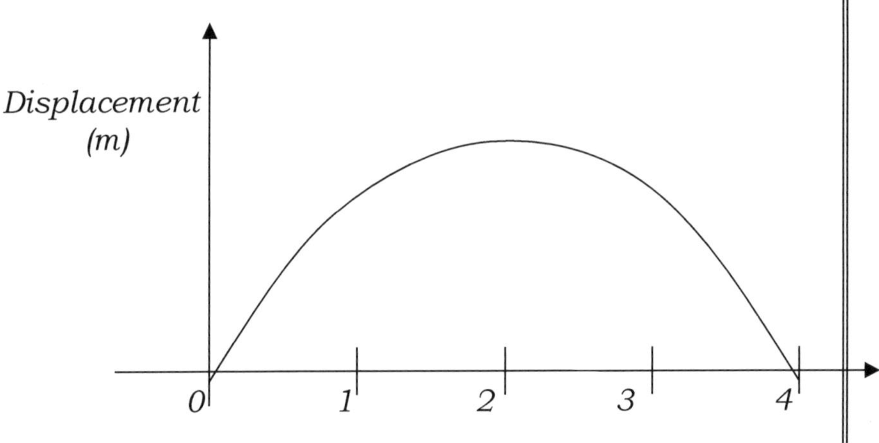

Time (s)

c. While maintaining a constant angular velocity, ω, the length of

horizontal section, r, of the string is measured for various values

of m attached to the vertical end of the string.

9. A new train is being tested along a straight track. Its speed is recorded every 20 seconds. The results are as shown in the velocity-time graph.

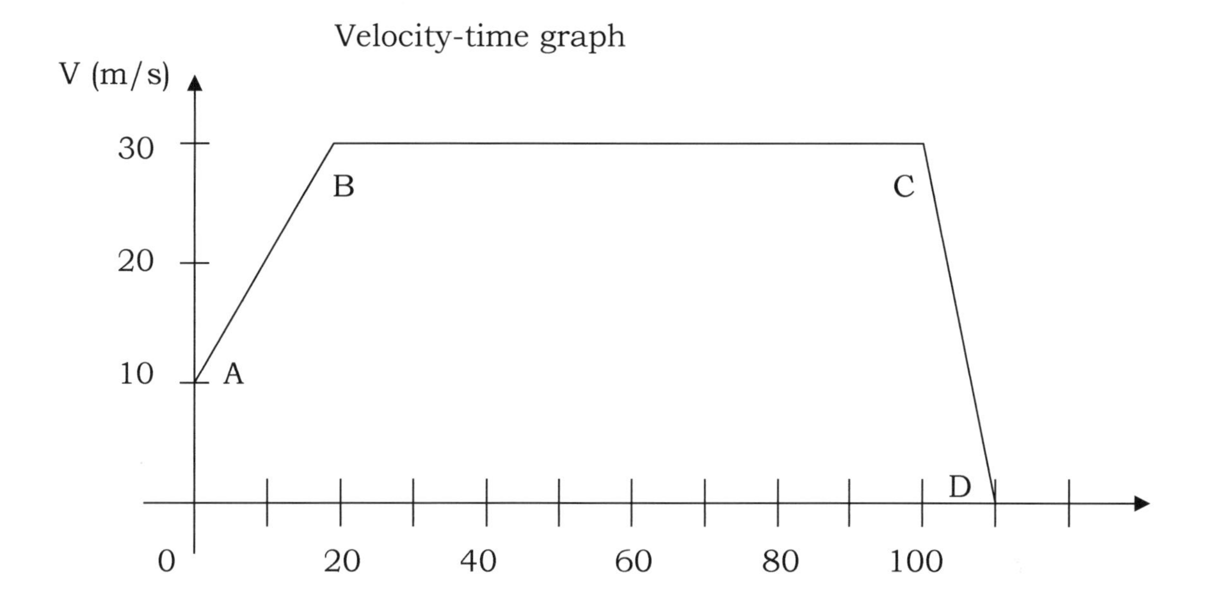

t (s)

a. Describe the motion of the train in section BC and CD.

b. How far does the train travel in part BC?

c. Calculate the acceleration of the train in the first 20 seconds.

Solution:

a. *BC* - *The train moves with a constant (uniform) velocity.*

CD - *The train decelerates/retards uniformly to rest.*

b. *Distance = Area under the graph*

= 30 x (110 – 20) = 2,700m

c. *Acceleration = gradient of the graph*

$= \frac{30 - 10}{20 - 0} = 1 m/s^2$

10. A pistol of mass 500g fires a 20g bullet at a velocity of 200m/s. Find the recoil velocity of the pistol.

Solution:

$m_pv_p = m_bv_b$ $\quad$ $500 \times v_p = 20 \times 200$

$v_p = 4000/500$

$v_p = 8m/s$

11. The sketch shows a velocity-time graph for a body of mass 2kg.

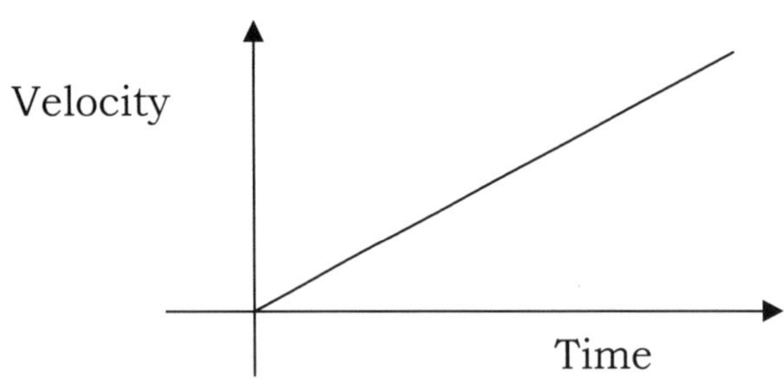

Given that there is only a single force acting on the body, sketch the corresponding:

a. Force-time graph for the above motion.

b. Displacement-time graph

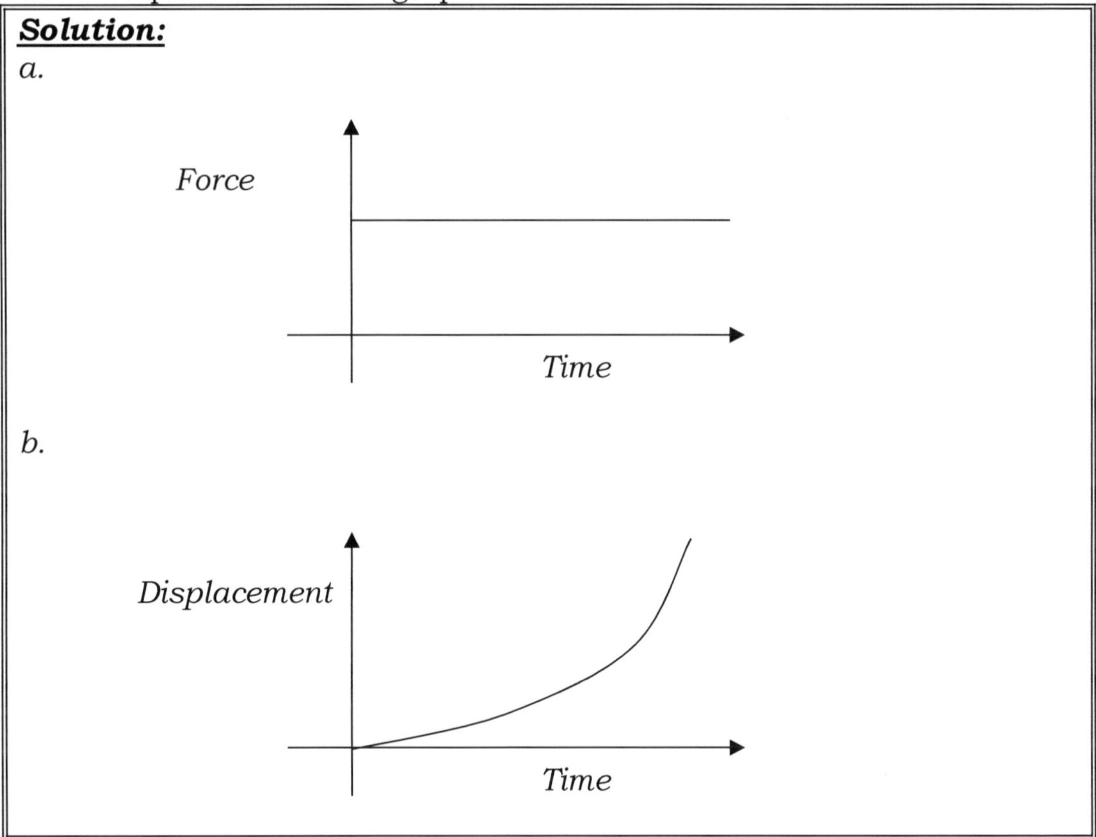

12. A person jumping from a 12m high building onto a concrete floor is likely to be seriously injured while the same person diving into a swimming pool would be quite safe. Explain.

(2 marks)

Solution:
On the floor, the person gets full reaction, while in water the uptrust reduces the reaction.

13. A car of mass 1,500kg moves round a bend of radius 40.5 metres on a flat road. If the coefficient of friction for the tyres and the road surface is 0.2, what is the maximum safe velocity at which the car can negotiate the bend? (Take g = 10N/kg)

Solution:

Centripetal force, $F = \frac{mv^2}{r} = \mu mg \quad \Leftrightarrow \quad v^2 = r\mu g$

$v^2 = 40.5 \times 0.2 \times 10 = 81$

$\therefore \quad v = 9m/s$

14. An ice hockey puck of mass 0.1kg travelling, at 20m/s is struck by a stick so as to return it along its original path 10m/s. Calculate the impulse of the force applied by the hockey stick.

Solution:

Impulse, $Ft = mv - mu$

$= m(v - u) = 0.1(20 - -10) = 0.1 \times 30$

$= 30Ns$

15. The figure shows a hand operated spray gun in use.

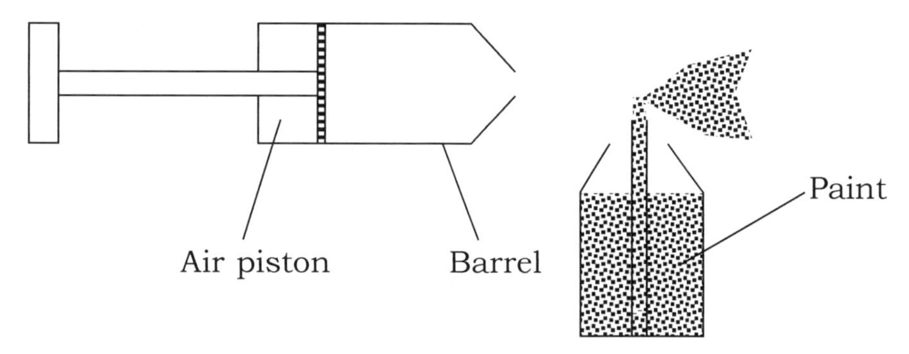

Explain how the gun functions

Solution:
The fast moving air from the barrel, above the vertical tube lowers pressure below atmospheric pressure. Hence the atmospheric pressure, just above the surface of the paint pushes it through the vertical tube, and is sprayed in a fine spray.

16. The figure shows two masses 0.3kg and 0.6kg connected by a light inextensible string through a tube.

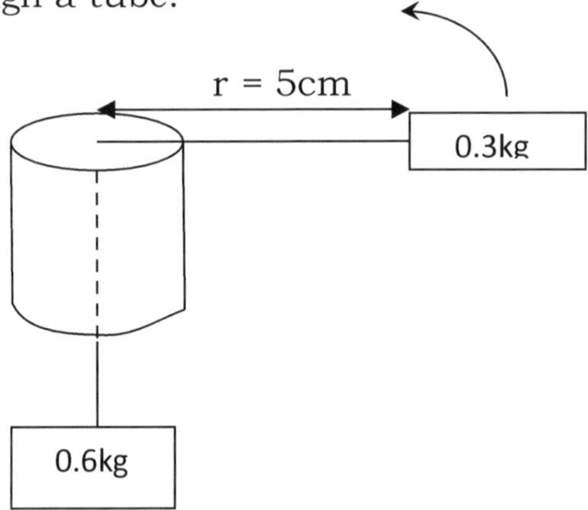

The 0.3kg mass is made to rotate in a horizontal circle of radius 5cm. Calculate the angular velocity of the mass when the system is in equilibrium. (Take g = $10m/s^2$)

Solution:

$m_1 g = m_2 \omega^2 r$

$\Rightarrow \quad 0.6 \times 10 = 0.3 \times \omega^2 \times 0.05 \quad \Leftrightarrow \quad \omega^2 = \frac{6}{0.015} = 400$

$\therefore \quad \omega = 20 rad/s$

17. The figure shows a part of a tape pulled through a ticker-timer by a trolley moving down an inclined plane.

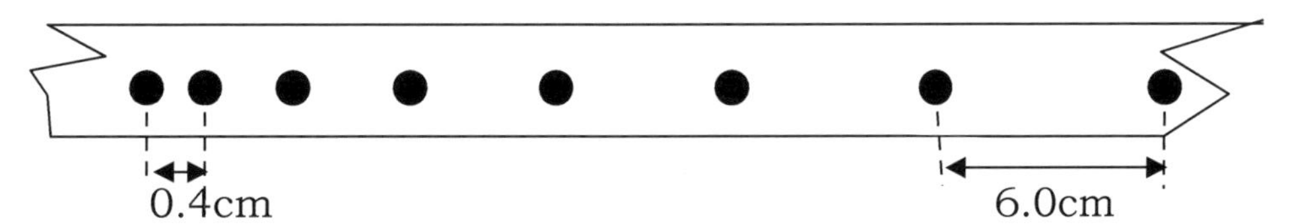

If the frequency of the ticker-timer is 50Hz, calculate the acceleration of the trolley.

Solution:

$v_1 = \frac{0.004}{0.02}$

$= 1/5 = 0.2m/s$

$v_2 = \frac{0.006}{0.02}$

$= 3m/s$

$\Leftrightarrow \quad a = \frac{3 - 0.2}{0.14}$

$= \frac{2.8}{0.14}$

$= 20m/s^2$

18. A bullet travelling at 300m/s strikes a thick wall and escapes at the other side of the wall in 0.005 seconds at a velocity of 100m/s. What is the thickness of the wall?

Solution:

From $v^2 = u^2 + 2aS \quad \Leftrightarrow \quad S = \frac{v^2 - u^2}{a}$

but $a = \frac{300 - 100}{0.005} = -40{,}000m/s^2$

$\therefore \quad S = \frac{100^2 - 300^2}{-40{,}000} = \frac{-80{,}000}{-40{,}000}$

$S = 2.0m$

19. Find the work done when a body of mass 50kg is dragged a distance of 40m with a constant velocity along a horizontal ground, if the coefficient of friction between the body and the floor is 0.40.

Solution:

Frictional force overcome, $F = \mu R$

$\Rightarrow$ $F = 0.4 \times 500 = 200N$

And Work = force x distance

$= 200 \times 40$

$= 8,000J$

20. In a dormitory in a certain school, showers are fixed on a certain pipe of varying diameter as shown in the figure.

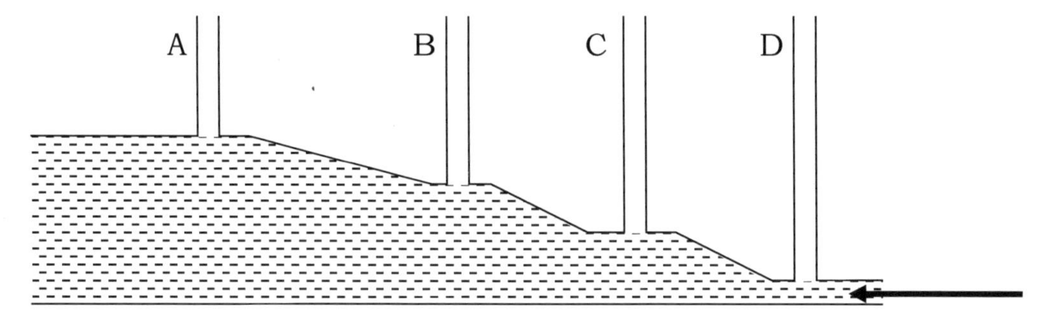

If water travels in the direction shown, which one of the showers at A, B, C or D is most likely to have running water even when the pressure in the main pipe becomes low? Explain.

Solution:

Shower A. The pressure is high because the velocity of water is low.

21. A boy of mass 50kg stands on a scale balance in a lift, which is accelerating upwards. The acceleration is $2.0ms^{-2}$ at one instant. Calculate the reading on the scale at that instant.

Solution:

Downward force $= mg = 50 \times 10 = 500N$

Upward force $= ma = 50 \times 2 = 100N$

Resultant force $= 500 - -100 = 600N$

$\therefore$ *Reading on scale = 600N Or 60kg.*

22. The figure shows a tall measuring cylinder containing a viscous liquid. A very small steel ball is released from rest at the surface of the liquid as shown.

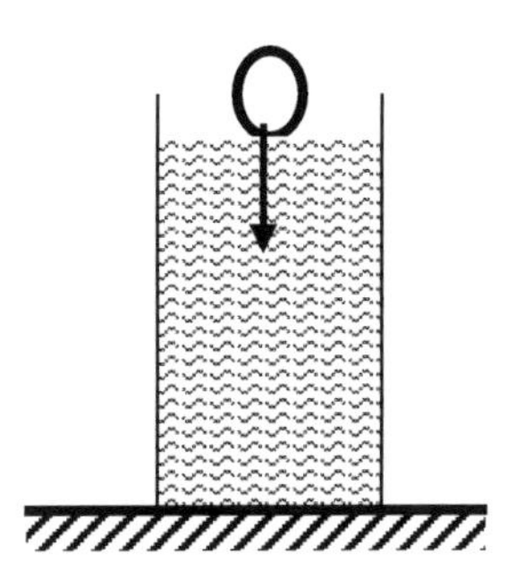

Sketch the velocity-time graph for the motion of the ball from the time it was released to the time just before reaching the bottom of the cylinder.

Hence explain your graph.

Solution:

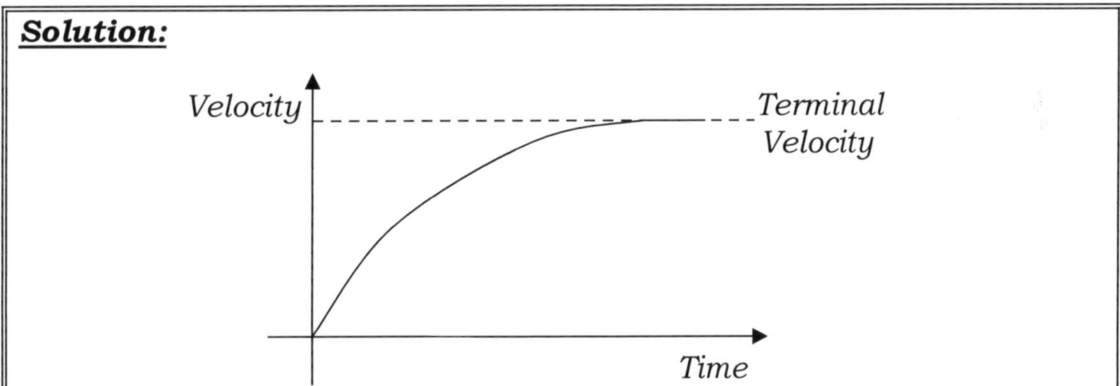

When the ball enters into the liquid, its weight is greater than the upthrust and the viscous drag; hence the resultant downward force increases its velocity. Later the upward and downward forces become equal and the ball attains a constant velocity, the terminal velocity.

23. A trolley of mass 40kg is initially at rest on a horizontal surface. It is connected by a light inextensible rope running over a frictionless pulley to a mass of 10kg.

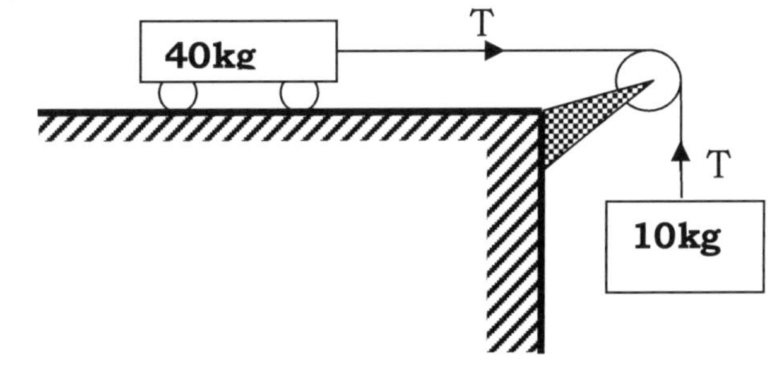

What is the acceleration of the masses when the system starts to move?

Solution:

$$\left.\begin{aligned} 100 - T &= 10a \quad \ldots\ldots\ldots\ldots\ldots (i) \\ T &= 40a \quad \ldots\ldots\ldots\ldots\ldots (ii) \end{aligned}\right\} \Rightarrow \quad 100 = 50a \quad \therefore \quad a = 2m/s^2$$

24. Water flows trough along a horizontal pipe of cross sectional area of 48cm^2 at one instant. If the speed of the water at a constriction of cross sectional area of 12cm^2 is 4m/s, calculate the speed of the water in the wide section.

Solution:

$v_1A_1 = v_2A_2$ *(Equation of continuity)*

$\Rightarrow \quad 48v_1 = 12 \times 4 \quad \Leftrightarrow \quad v_1 = \dfrac{12 \times 4}{48}$

$\therefore \quad v_1 = 1m/s$

25. Explain how Bernoulli's theorem and the law of continuity are applied in the working of a Bunsen burner.

Solution:

- *Gas opens, the jet jets out gas at high velocity decreasing the pressure.*
- *High pressure outside forces air into the barrel hence mixing the gas and air for burning.*

26. A hot air balloon is tethered to the ground on a windless day. The envelope of the balloon contains 1,200m^3 of hot air of density 0.8Kg/m^3. The mass of the balloon (not including the hot air) is 400Kg. If the density of the surrounding air is 1.3Kg/m^3, calculate the tension in the rope holding the balloon on the ground.

Solution:

Upward force = uptrust = weight of air displaced

$= 1{,}200m^3 \times 1.3kg/m^3 \times 10N/Kg$

$= 15{,}600N$

Downward forces = Tension + Weight of hot air + weight of fabrics

$= T + (1{,}200 \times 0.8 \times 10) + (400 \times 10)$

$= T + 9{,}600 + 4{,}000$

$= T + 13{,}600$

Since Upward forces = downward forces

$\Rightarrow \quad 15{,}600N = (T + 13{,}600)N$

$\Leftrightarrow \quad T = 15{,}600N - 13{,}600N$

$\therefore \quad T = 2{,}000N$

27. If a stone is dropped from rest down a well and water splashes 2.5 seconds later, how deep is the well?

Solution:

$S = {}^1/_2 gt^2 = {}^1/_2 \times 10 \times 2.5 \times 2.5$

$= 31.25m$

28. A mass of 8Kg is whirled round in a vertical circle using a rope of length 80cm. If it makes 2.5 circles in 1s, calculate the maximum tension the rope experiences.

Solution:

For maximum tension, $T = F + mg$ *where F is the centripetal force.*

$\Leftrightarrow \quad T = m\omega^2 r + mg$

but $\omega = 2.5rev/s = 2.5 \times 2 \times 3.14 = 15.7rad/s$

$\therefore \quad T = 8 \times (15.7)^2 \times 0.8 + 8 \times 10$

$= 1{,}577.54 + 80 = 1{,}657.54N$

Chapter 11: Magnetism and Electromagnetic Induction

1. The figure represents a step-down with 500 turns in the primary and 50 turns in the secondary. The turns are wound uniformly on the core. The lengths of PQ and QR are as indicated.

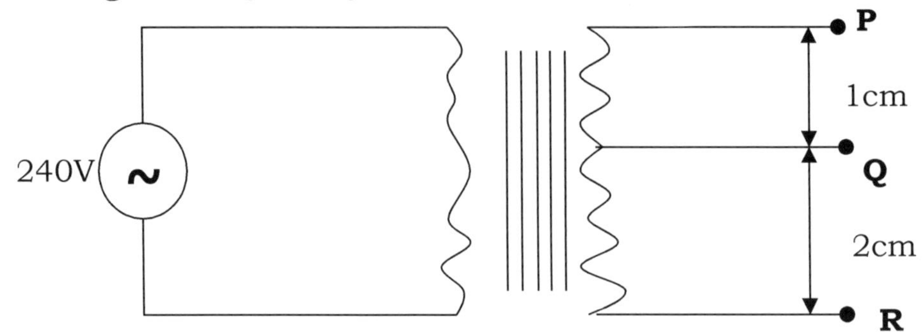

Determine the p.d. across PQ.

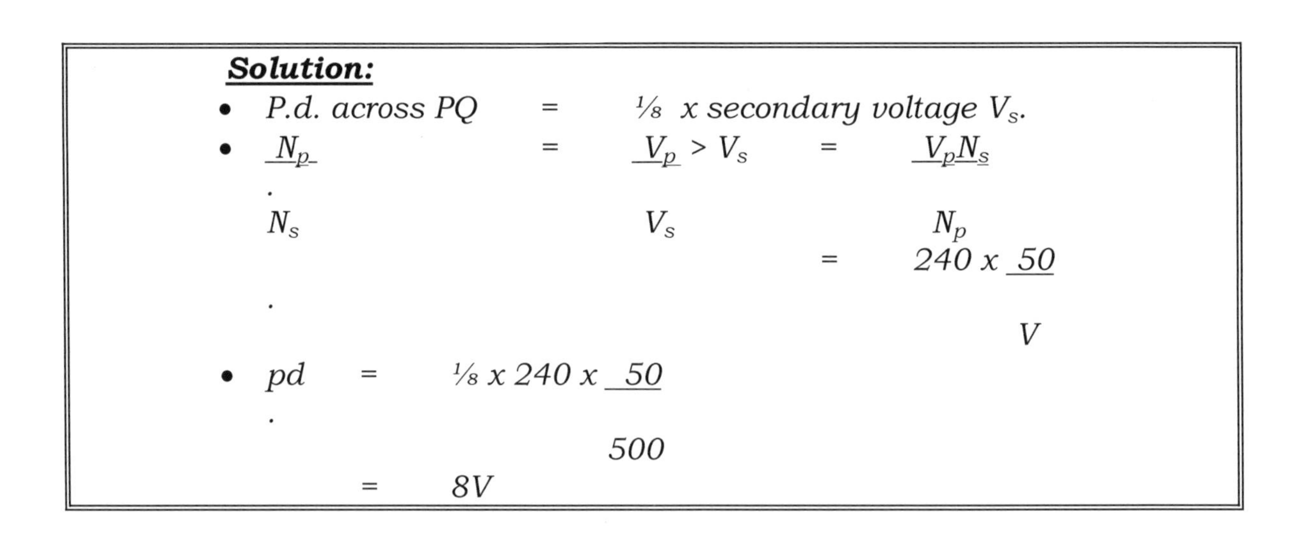

Solution:

- *P.d. across PQ* = $\frac{1}{8}$ *x secondary voltage* V_s.
- $\frac{N_p}{N_s} = \frac{V_p}{V_s}$ > $V_s = \frac{V_p N_s}{N_p}$
  $= 240 \times \frac{50}{V}$
- *pd* = $\frac{1}{8} \times 240 \times \frac{50}{500}$
  = *8V*

2. a. The primary coil of a transformer draws a current of 50mA when connected to a 240V a.c. supply. The secondary coil is connected to a 5Ω resistor has a current of 1.4A flowing through it. Calculate:

   i. The power supplied to the transformer.

   ii. The efficiency of the transformer.

   b. State two ways in which to energy losses in a transformer are minimized.

Solution:

a. i. *$P = IV = 240V \times 0.05A = 12$ watts.*

ii. *Efficiency* $= \frac{\text{Power output}}{\text{Power input}} \times 100\%$

$= \frac{I2R}{12W} \times 100\%$

$= \frac{1.4^2 \times 5\Omega}{12} \times 100\%$

$= \frac{9.8}{12} \times 100\%$

$= 81.7\%$

b. i. *Laminating the core.*

ii. *Use of thick copper wire*

iii. *Use of common core*

iv. *Use of soft iron core.*

3. a. A coil of an insulated copper wire is wound round a U-shaped soft iron

core and connected to a battery as shown in the figure below.

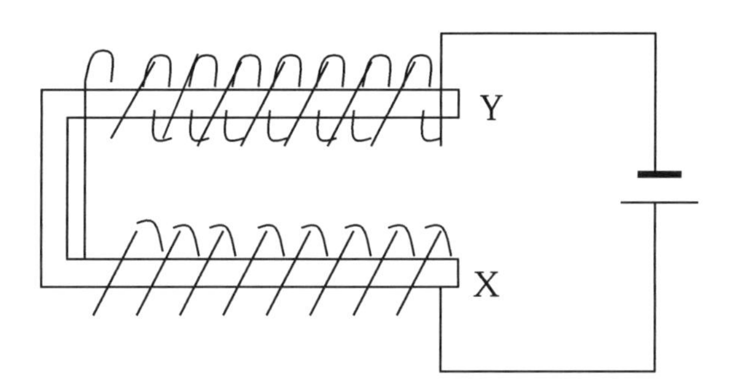

Indicate on the diagram by use of an arrow the direction of current and state the polarities at X and Y.

b. A bar magnet was thrashed into a coil connected to a centre zero galvanometer as show in the diagram below. Indicate the direction of the induced current.

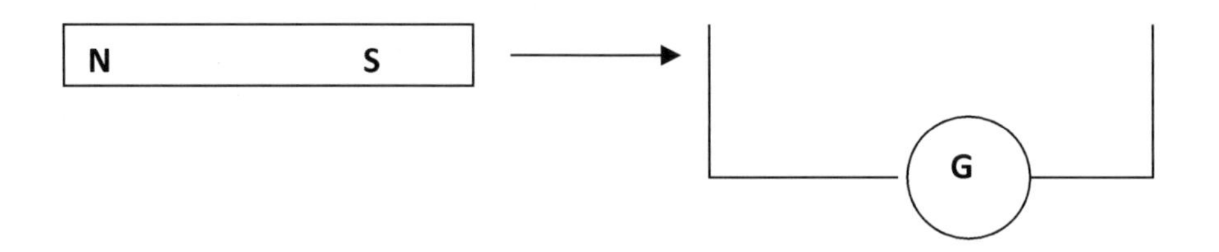

c. i. A transformer must use a.c. current and not d.c. Explain why.

ii. State four ways in which a transformer may lose energy.

iii. A 12V, 60Wheater can be operated by a 24V main transformer,
which has 10,000 primary turns. If the transformer is 100% efficient, calculate

- The number of secondary turns.
- The current flowing in the primary coil.

Solutions:

a.

b.

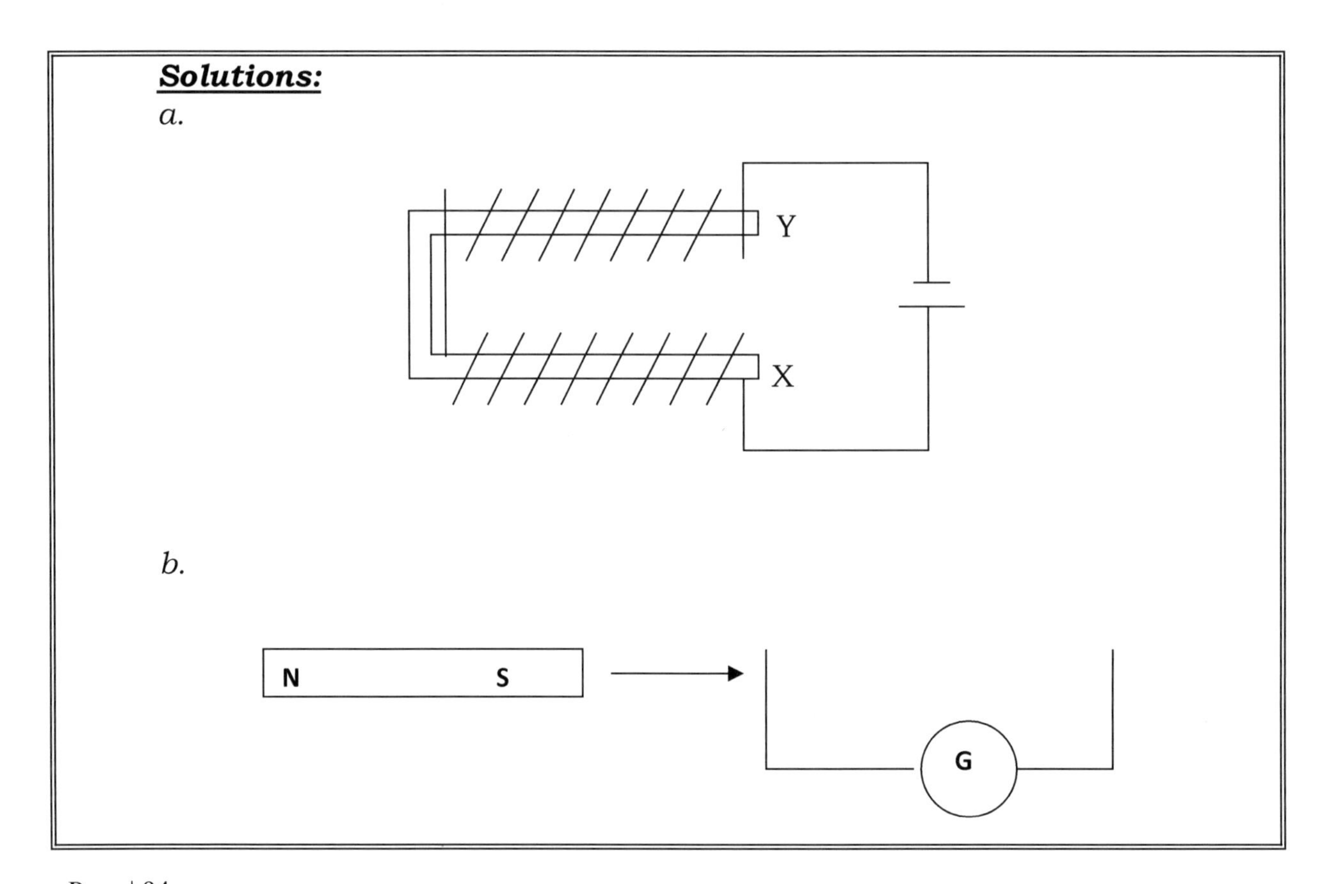

c. i. *A.C. varies therefore will produce the change in magnetic flux needed for induction while D.C. does not vary, no induction.*

ii.
- *Energy losses due to resistance of windings.*
- *Eddy current losses*
- *Design of the coil*
- *Hysteresis losses*
- *Flux linkage.*

iii. I. $\frac{E_S}{E_P} = \frac{N_S}{N_P}$ $\qquad \frac{12}{240} = \frac{N_S}{10,000}$

$N_S = 500$

II. $I_P V_P = 60 \qquad I_P = \frac{60}{240}$

$= 0.25A$

4. The diagram shows a step-up transformer commonly used at power stations.

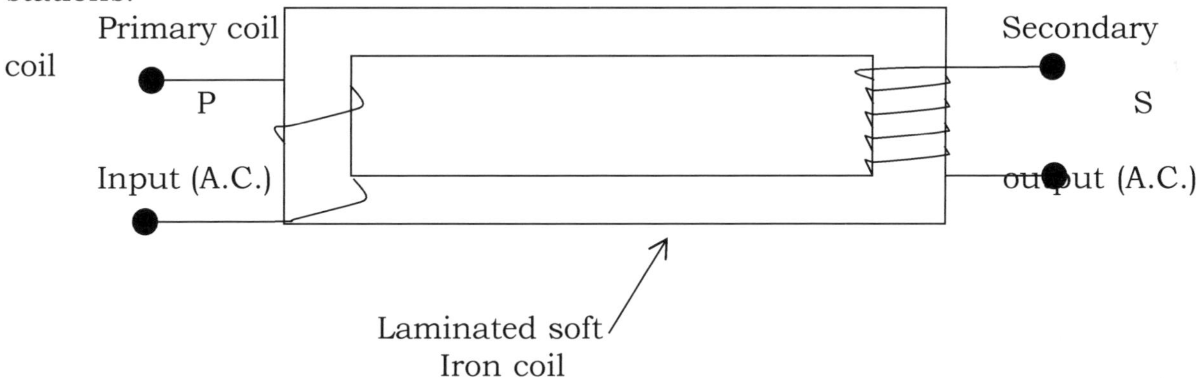

a. What is meant by a step-up transformer?

b. How do you determine that the above transformer is a step-up one?

c. What is the purpose of the 'soft iron core'?

d. Why is the e.m.f. produced at a power station stepped up to a high voltage for long distance transmission.

e. Calculators work on about 6V d.c. Explain how they are adopted to work by drawing power from the mains.

f. An induction coil is in fact, a step-up transformer designed to be used with d.c. Explain how the changing magnetic field is produced using a d.c. in the primary coil.

Solution:

a. A step-up transformer changes primary alternating voltage to a higher value.

b. The secondary coil has more turns than those in te primary coil.

c. The soft iron is easily magnetised and demagnetised. When magnetised, it becomes a strong electromagnet and increases the strength of the magnetic filed.

d. The power loss in transmission lines is due to the heating effect of electricity in the cable. Since heat (current^2 x resistance) is produced, the power loss through heat can be minimised by keeping both the current and resistance of the cable as low as possible. The resistance of the wire is kept low by using thick wires with large cross-sectional area. Since power = current x voltage, the same power can be carried in a cable at low current – if the voltage is made high. A low current will reduce the power loss, as heat produced will be less.

e. The mains is 240V. The calculator is operated using an adaptor, which reduces the main voltage to 6V through a step-down transformer. The a.c. is then rectified to d.c. by a rectifier before supplying to the calculator.

f. The primary current is d.c. A 'make and break' device used in the primary circuit continually closes and opens the circuit at a very rapid rate. This result in the rapid changing of the magnetic flux associated with the coils. Hence an induced e.m.f is obtained across the terminals of the secondary coil.

5. Explain why a freely suspended magnet always comes to rest in a north-south direction.

Solution:

This is due to attraction between unlike poles. The earth's magnetic north pole is in the geographic South Pole and its magnetic south pole is in the North Pole.

6. An iron nail is being magnetised by single touch method as shown in the figure.

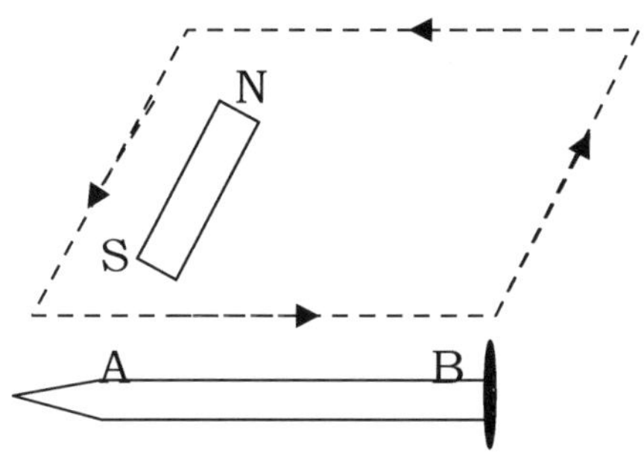

State the polarities of A and B, giving a reason.

> ***Solution:***
> *A – South Pole and B – North Pole.*
> *Te polarity produced at that end of the bar where stroking finishes is opposite to that of the stroking pole.*

7. State what should be done in order to change the direction of rotation of coil of a D.C. motor.

> ***Solution:***
> - *Change the direction of current by interchanging terminals.*
> - *Change the direction of the magnetic field by interchanging poles of the magnets.*

8. A rectangular coil connected to a galvanometer is placed in a magnetic field as shown in the figure.

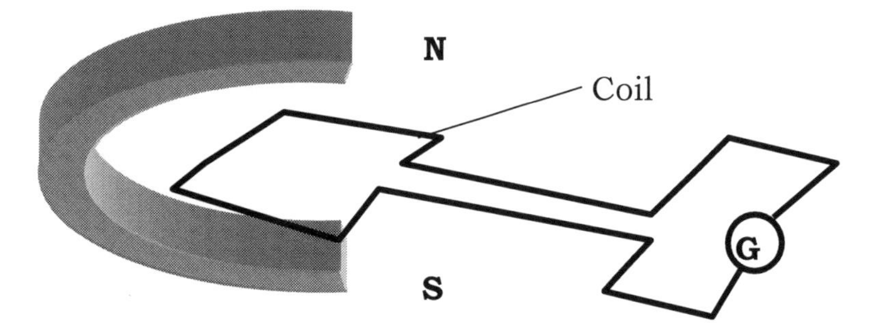

Explain the observation that, 'when the coil is held stationary in the magnet, no deflection is noted but when the coil is pulled outwards a deflection is noted'.

Solution:

When held stationary, there is no change in magnetic flux linkage, no induced e.m.f. As it is moved, change in flux linkage occurs causing an induced e.m.f.

9. A transformer with 1,000 turns in the primary coil is connected to a 400 volts a.c. source. The voltage across the secondary coil is 240 volts. Calculate the number of turns in the secondary coil.

Solution:

Turns ratio $\frac{N_P}{N_S} = \frac{V_P}{V_S} = \frac{I_S}{I_P}$

$\Rightarrow \frac{1,000}{N_S} = \frac{400}{240} \quad \Leftrightarrow \quad N_S = \frac{1,000 \times 240}{400}$

$\therefore N_S = 600$ *turns*

10. Explain what is meant by magnetic field.

Solution:

This is a region in which a magnetic force can be detected.

11. Sketch the magnetic field pattern between the two poles of the magnets shown in the diagram. The current carrying wire is in between the poles.

N  S

Solution:

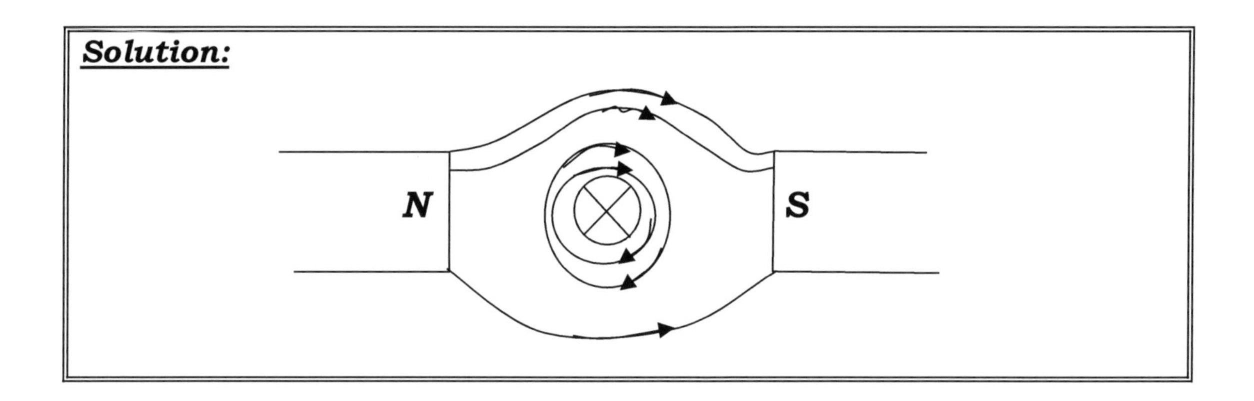

Chapter 12: Mechanical properties of matter

1. State the Hooke's law.

> ***Solution:***
> *Provided the elastic limit is not exceeded, the extension of a spring is directly proportional to the stretching force.*

2. The breaking stress of a cable is $2.2 \times 10^8 N/m^2$. If the cable has a diameter of 0.14cm, calculate the maximum load that can be supported by the cable.

> ***Solution:***
>
> $$\text{Breaking stress} = \frac{Force}{Area} = 2.2 \times 10^8$$
>
> $$\Rightarrow \quad \frac{F}{22/7 \times 0.07 \times 0.07} = 2.2 \times 10^8$$
>
> $$\Leftrightarrow \quad F = 2.2 \times 10^8 \times 22/7 \times 7 \times 7 \times 10^{-8}$$
>
> $$\therefore \quad F = 338.8N$$

3. Draw the level of mercury in the capillary tube shown in the figure.

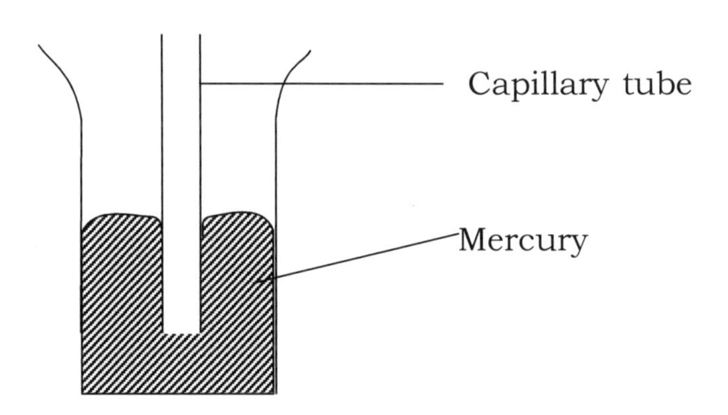

Solution:

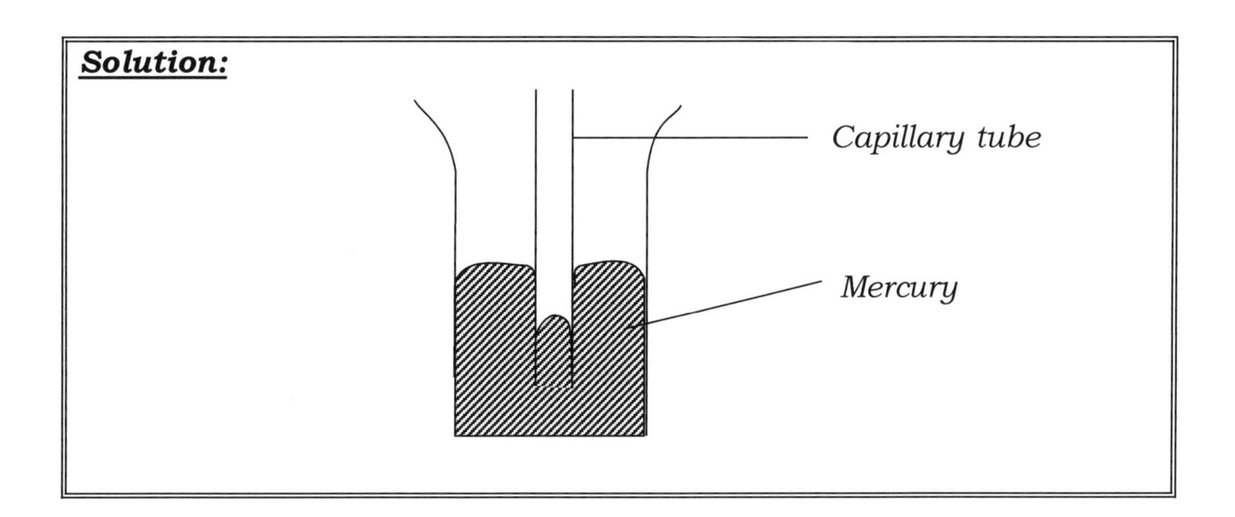

Chapter 13: Moments of Force

1. a. State the principle of moments.

   b. A non-uniform plank of wood MN is 2.50m long and weighs 950N. Spring balances X and Y are attached to the plank at a distance of 0.40m from each end as shown in the figure below.

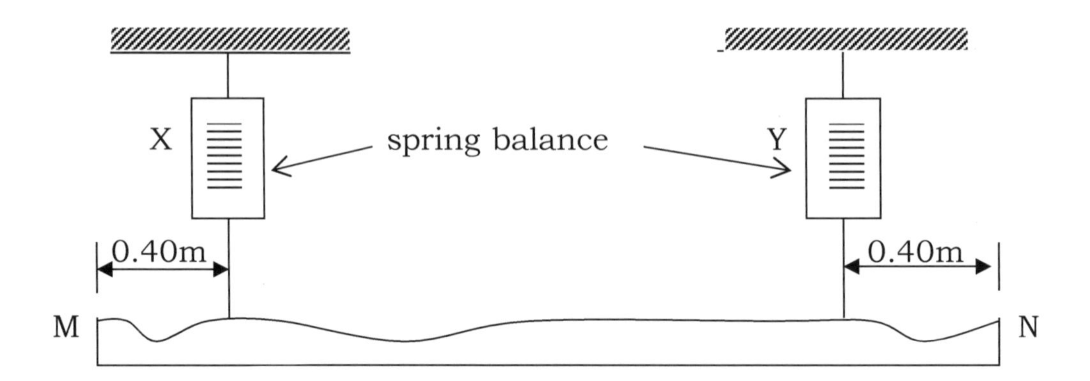

   When the plank is horizontal, balance X records 570N. Calculate the reading of balance Y.

   c. Determine the distance of the centre of gravity from the end 'M' of the plank.

Solution:

a. *For a system in equilibrium, the sum of the clockwise moments about a point is equal to the sum of the anticlockwise moment about the same point.*

b.

$W = X + Y$

$\Rightarrow \quad 950 = 570 + Y$

$\Leftrightarrow \quad Y = 380N$

c. *Taking moments about M*

$\Rightarrow \quad 570 \times 0.4 + 380 \times 2.1 = 950 \times d$

$\Leftrightarrow \quad 950d = 228 + 798$

$\Leftrightarrow \quad d = \frac{1,026}{950}$

$d = 1.08m$

2. Two masses X and Y each of mass 3kg balances on light beam 5m long when Y is fully immersed in a liquid of density $2g/cm^3$.

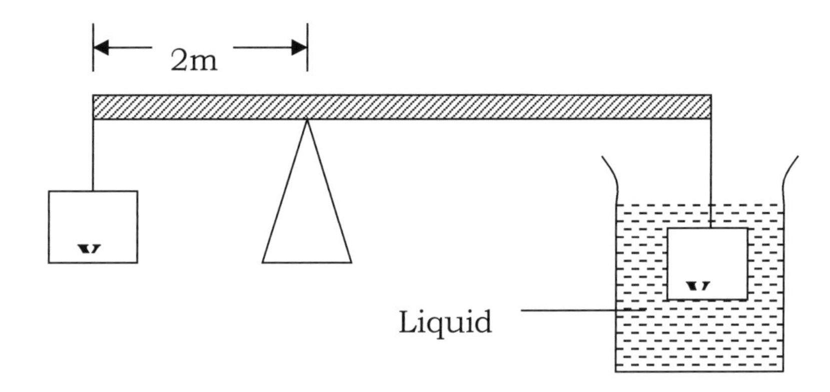

a. Calculate the upthrust on mass Y.

b. What is the volume of the liquid displaced by Y?

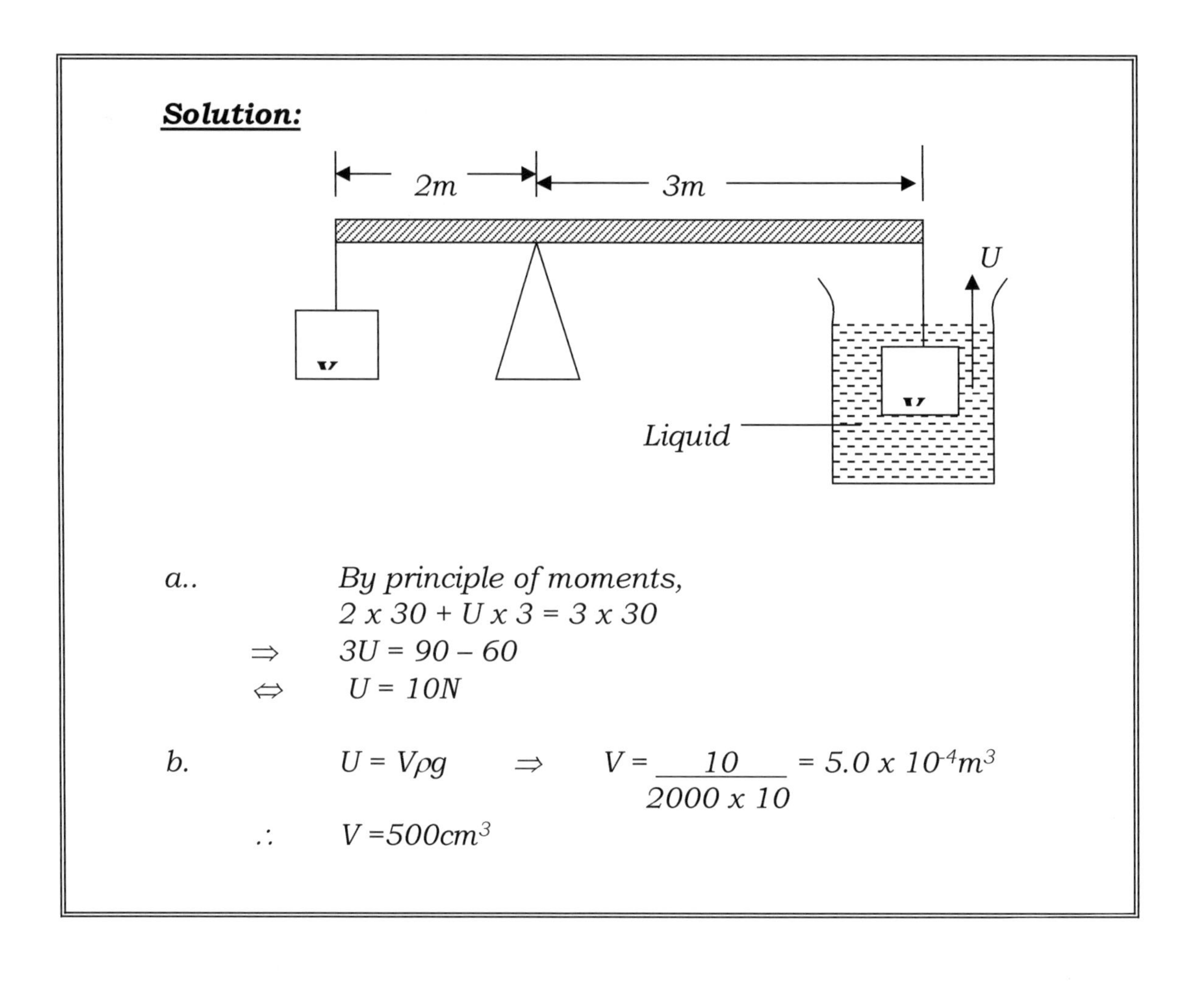

3.

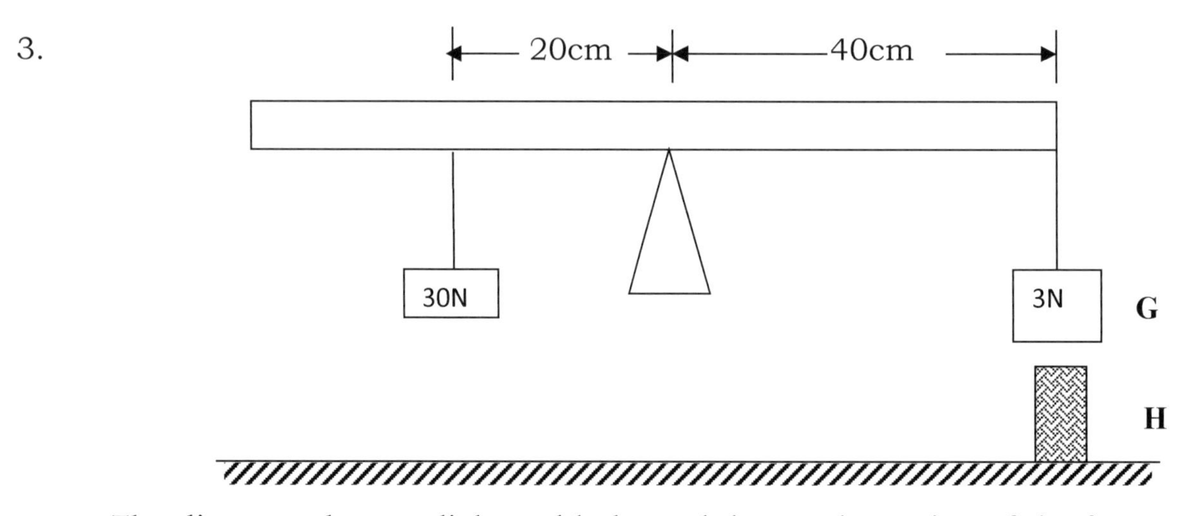

The diagram shows a light rod balanced due to the action of the forces shown. G is a magnet of weight 3N and H is a permanent magnet. Determine the force between G and H stating whether it is attractive or repulsive.

Solution:

Clockwise moments = anticlockwise moments

$30 \times 20 = 3 \times 40 + X \times 40 \quad \Leftrightarrow \quad 600 = 120 + 40x$

$40x = 480 \quad \therefore \quad x = 12N$

N.B. The force is attractive since it is positive; to the direction of the 3N force.

4. A uniform meter-rule of mass 300g is pivoted freely at the 0cm mark as shown in the diagram.

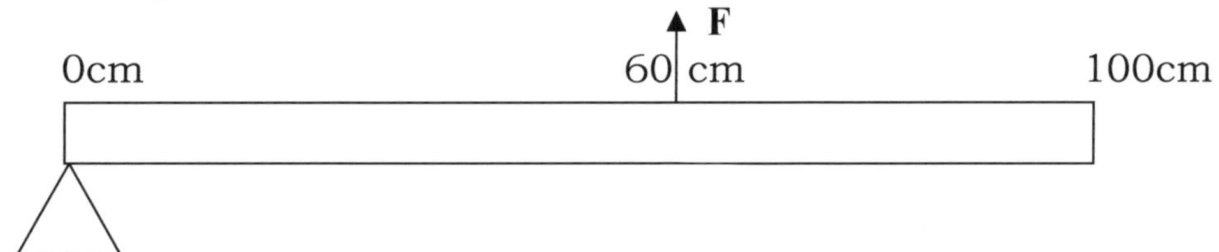

What force applied vertically upwards at the 60cm mark is needed to maintain the rule horizontally?

Solution:

By the principle of moments, clockwise moments = anticlockwise moments.

$\Leftrightarrow \quad 60 \times F = 3 \times 50$ *N.B. The weight of the rule* $\frac{300 \times 10}{1,000}$ *N is at its centre.*

$\Leftrightarrow \quad F = 2.5N$

5. A uniform bar 100cm long is pivoted 25cm from one end. A 10N weight hangs 5cm from the same end and keeps the system in equilibrium. Determine the weight of the bar.

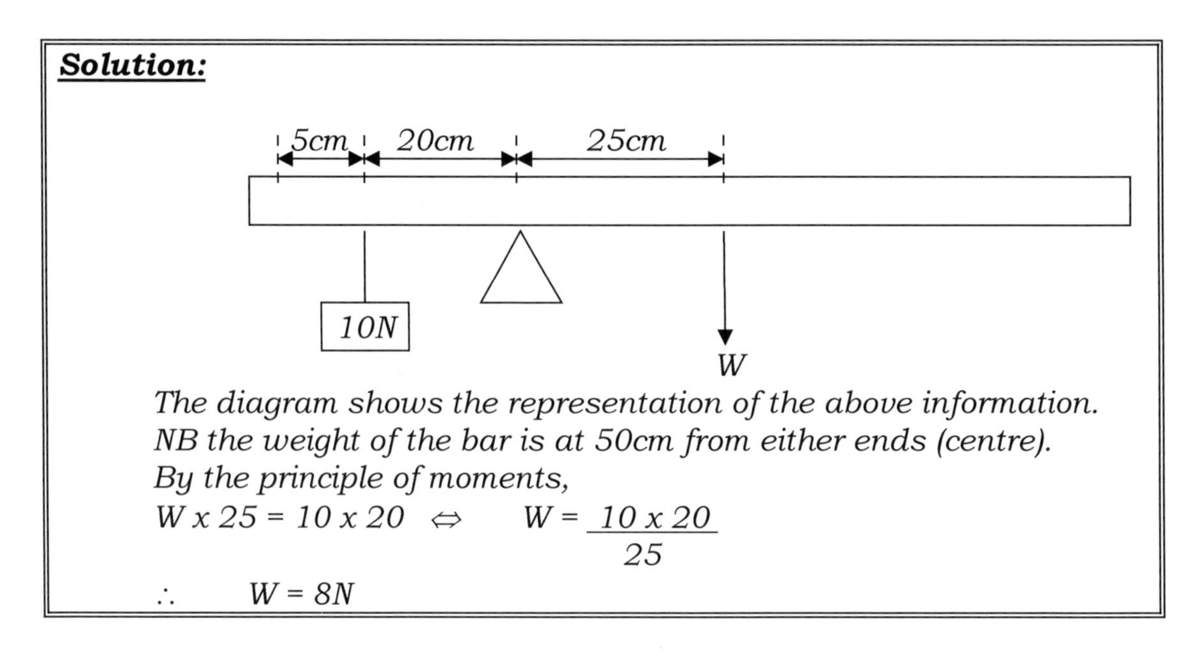

Solution:

The diagram shows the representation of the above information.
NB the weight of the bar is at 50cm from either ends (centre).
By the principle of moments,

$W \times 25 = 10 \times 20 \Leftrightarrow W = \frac{10 \times 20}{25}$

$\therefore \quad W = 8N$

6. The figure shows a uniform meter rule balanced by two forces A and B.

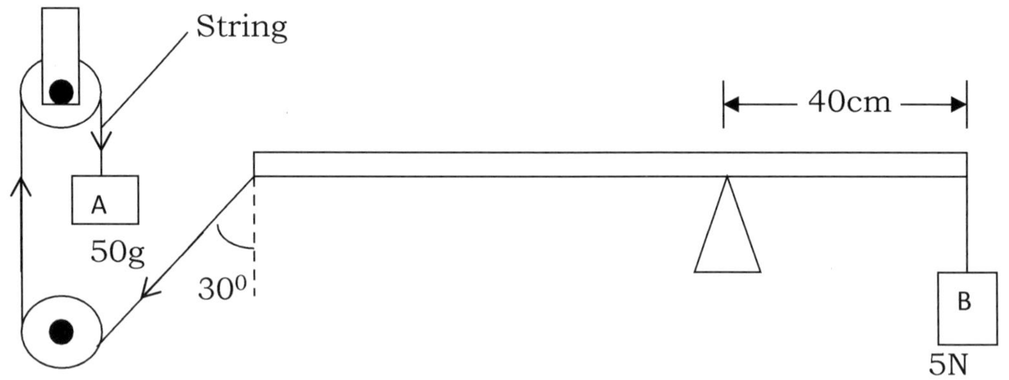

Assuming there is mo frictional force on the fixed pulley; calculate the weight of the meter rule.

Solution:

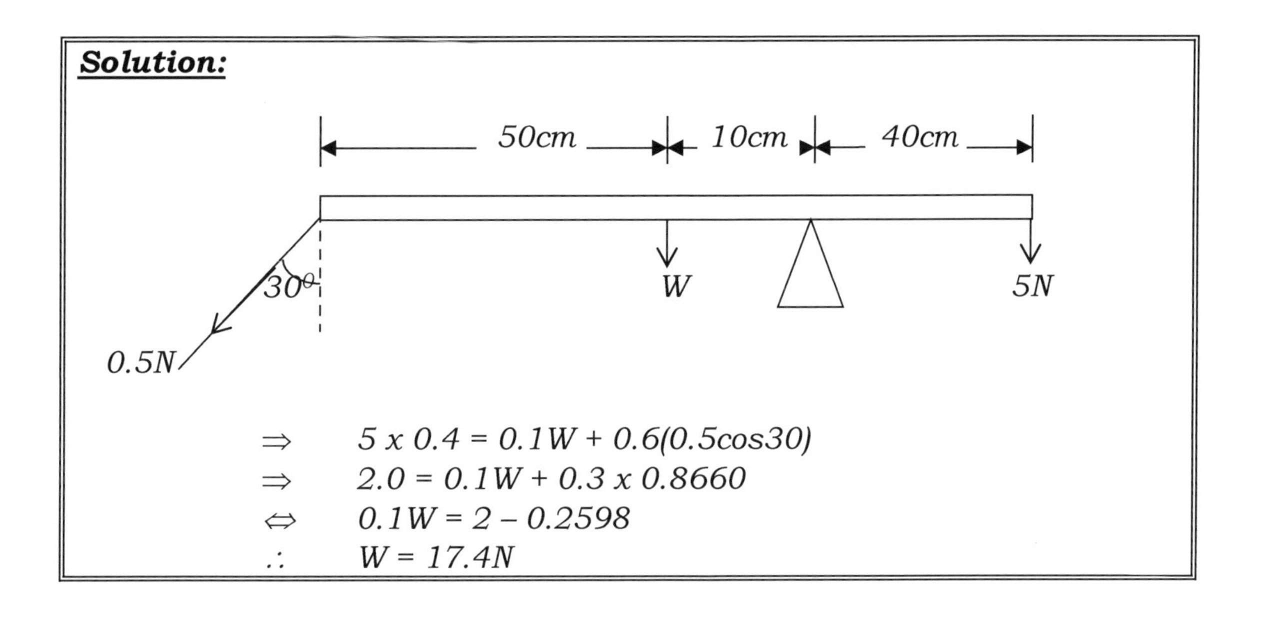

$\Rightarrow \quad 5 \times 0.4 = 0.1W + 0.6(0.5\cos 30)$

$\Rightarrow \quad 2.0 = 0.1W + 0.3 \times 0.8660$

$\Leftrightarrow \quad 0.1W = 2 - 0.2598$

$\therefore \quad W = 17.4N$

7. A uniform meter rule is suspended vertically from a pivot at the 0cm mark as shown in the figure. Given that the 24N force acts at the 10cm mark while the 16N force acts at the 90cm mark, calculate the force F, which acts at the 60cm mark.

Solution:

$\Rightarrow \quad F \times 0.6 = 24 \times 0.1 + 16 \times 0.9$

$\Leftrightarrow \quad 0.6F = 16.8$

$\therefore \quad F = 28N$

8. The figure shows a uniform bar pivoted at its centre and in equilibrium under the forces shown.

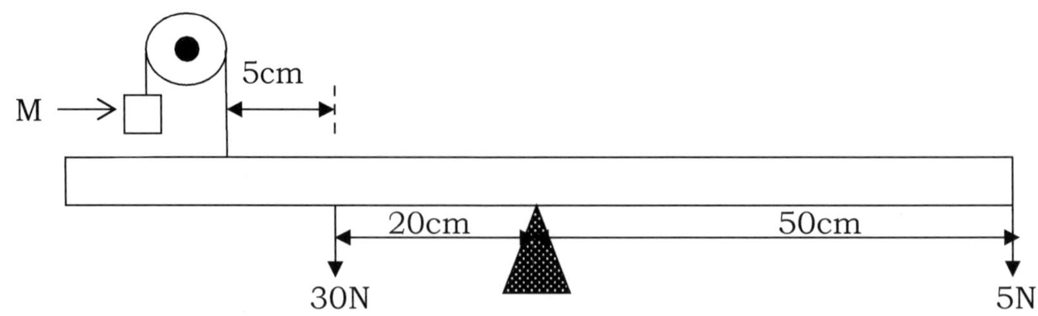

Determine the mass M.

Solution:

Clockwise moments = Anti-clockwise moments

$\Rightarrow$ $mg \times 25 + 5 \times 50 = 30 \times 20$

$\Leftrightarrow$ $10m \times 25 + 250 = 600$

$\Leftrightarrow$ $250m = 350 \therefore \quad m = \frac{350}{250}$

$m = 1.4Kg$

Chapter 14: Photo Electric Effect

1. A photon has energy of 5×10^{-19}J. Calculate the wavelength associated with this photon. (Take the speed of light = $3.0 \times 10^{8}ms^{-1}$)

> ***Solution:***
>
> *Energy in a photon,* $E = \frac{hc}{\lambda}$
>
> $\therefore \quad \lambda = \frac{hc}{E} = \frac{6.63 \times 10^{-34} \times 3 \times 10^{8}}{5 \times 10^{-19}}$
>
> $\lambda = 3.978 \times 10^{-7}m$

2. The work function of a particular substance is 3.6eV. Determine the threshold frequency fro the substance. (1ev = 1.6×10^{-19}J and Planks constant, h = 6.62×10^{-34}Js)

> ***Solution:***
>
> $W_0 = hf_0$ *where* w_0 *= work function and* f_0 *= Threshold frequency.*
>
> $\Rightarrow \quad 3.6 \times 1.6 \times 10^{-19} = 6.62 \times 10^{-34} \times f_0$
>
> $\Leftrightarrow \quad f_0 = \frac{5.76 \times 10^{-19}}{6.62 \times 10^{-34}}$
>
> $= 8.7 \times 10^{14}Hz$

3. A high intensity ultra violet radiation is made to fall onto a clean zinc plate connected to a sensitive milliammeter as shown in the figure.

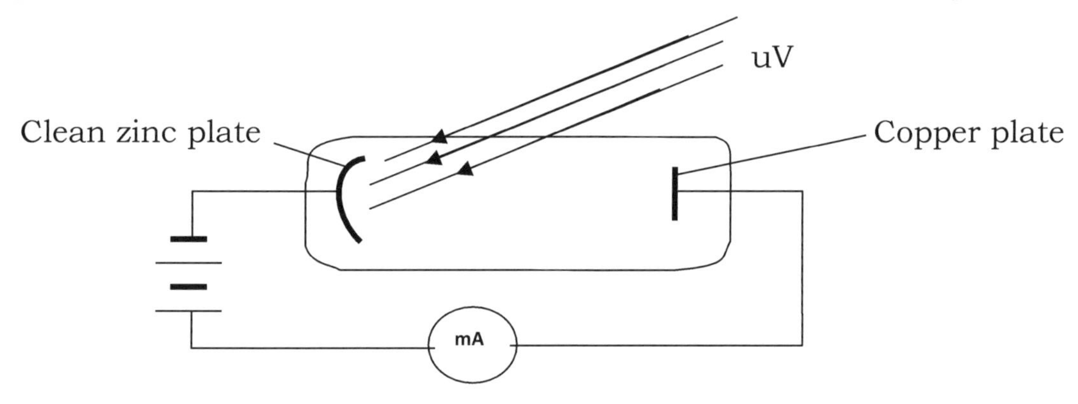

State and explain the observation made.

Solution:

- *Milliameter deflects.*
- *When uV falls on zinc, it ejects electrons from its surface. They move towards the Anode causing current flow.*

4. State two factors that affect photoelectric effect.

Solution:

- *Intensity of the radiation.*
- *Energy of the radiation.*

Chapter 15: Quantity of Heat

1. An engine of certain motorcar has an input power of 240Kw of which 75% is wasted as heat energy. This energy is carried away from the engine by water, which is cooled as it passes through the radiator of the car. The cooling system contains 6kg of water.
   a. How much heat energy must be carried away from the engine by the water every second.

   b. The temperature of the water rises from 55^0C to 95^0C while passing the engine. How much heat energy does the water absorb in the process.

   c. In how many seconds will the water pass through the engine so that it can carry that heat energy?

Solution:

a. *Heat carried away per second is 75% of 240kw*

$\therefore$ *75% x 240,000 = 180,000J/s*

b. *Heat absorbed* $= mc\Delta\theta$

$= 6 \times 4{,}200 \times (95 - 55)$

$= 108{,}000J$

c. *180,000J* $\Leftrightarrow$ *1 sec*

$\therefore$ *108,000J* $\Leftrightarrow$ $\dfrac{108{,}000}{180.000} = 5.6$ *sec.*

2. A girl dips one finger in water and another in methylated spirit both at room temperature. She then raises both fingers in the air. Explain the feelings on the fingers.

Solution:

The finger with the methylated spirit feels cooler than the one with water. The spirit evaporates faster taking away the latent heat of vaporisation from the finger leaving it feeling cool.

3. A concrete wall stops a bullet of mass 0.80g, travelling at $400ms^{-1}$. Calculate the amount of heat energy transferred to the wall.

Solution:

Kinetic energy possessed by the bullet is converted to heat energy.

$\therefore$ *Heat energy* = *k.e.* = $\frac{1}{2}\ mv^2$

$= \frac{1}{2} \times 8 \times 10^{-4} \times (400)^2$

$= 64J$

4. a. Define the terms:
   i. Specific heat capacity
   ii. Specific latent heat of fusion.

   b. The melting point of a certain bullet is 300^0C. If the initial temperature of the bullet is 20^0C, determine the least speed at which the bullet would be moving so that it melts when suddenly stopped.
   (Specific heat capacity of bullet = $840Jkg^{-1}K^{-1}$ and specific latent heat of fusion of material making the bullet = $6.3 \times 10^4\ JK^{-1}$)

Solution:

a. i. Specific heat capacity is the heat required to raise the temperature of a unit mass of a substance by 1^0C or 1K.

ii. Specific latent of fusion is the heat required to completely melt a unit mass of ice without change in temperature.

b. Heat required to raise the temperature of the bullet from 20^0C – 300^0C and then melt it completely is the kinetic energy the bullet will be possessing.

$\therefore$ $k.e. = \frac{1}{2} mv^2 = mc\Delta\theta + ml$

$\Rightarrow$ $\frac{1}{2} mv^2 = 840 \times (300 - 20) + 6.3 \times 104$

$= (840 \times 280) + 63,000 = 264,600$

$\therefore$ $v^2 = 264,600 \times 2 = 529,200$

$\Leftrightarrow$ $v = 727.46\ m/s$

5. 3kg of hot water was added to 9kg of water at 10^0C and the resulting temperature was 20^0C. Ignoring heat gained by the container, determine the initial temperature of the hot water.

(Specific heat capacity of water = 4,200 $Jkg^{-1}K^{-1}$)

Solution:

Heat lost by hot water = heat gained by cold water

$\Rightarrow$ $mc(\theta - 20) = mc(20 - 10)$ *where θ is the initial temp. of hot water.*

$3(\theta - 20) = 9 \times 10$

$\therefore \quad \theta = 30 + 20 = 50^0C$

6. An energy saving stove when burning steadily has an efficiency of 80%. The stove melts 0.04kg of ice at 0^0C in 160 seconds.
(Latent heat of fusion of ice = $336Jg^{-1}$)
Calculate:
   a. The power of the stove.

   b. The heat energy wasted by the stove.

Solution:

a. $$\text{Power} = \frac{ml}{} = \frac{0.04kg \times 336 \times 1{,}000\ J/kg}{160\ sec}$$
$$= 84\ J/s\ (watts)$$
$\Rightarrow$ *80% power of stove = 84J*
$$\therefore \quad \text{power of stove} = \frac{84 \times 100}{80}$$
$$= 105\ watts$$

b. *Heat energy wasted* $= (105 - 84) \times 160\ J$
$= 21 \times 160J$
$= 3{,}360J$

7. a. Define latent heat of vaporisation.

   b. Explain why water does not get hot in the ocean yet the temperatures around are very high.

c. A piece of copper of mass 40g at 200^{0}C is placed in a copper calorimeter of mass 60g containing 50g of water at 10^{0}C. Ignoring heat losses, find the final temperature after stirring. (Take specific heat capacity of copper and water to be 4,00Jkg^{-1}K^{-1} and 4,200Jkg^{-1}K^{-1} respectively.)

Solution:

a. *Latent heat of vaporisation is the amount of heat required to change a unit mass of a substance from liquid to vapour without any change in temperature.*

b. *Water has a high specific heat capacity; therefore a lot of heat is needed for every unit mass of water.*

c. *Heat lost by copper = heat gained by water and calorimeter*

$\Rightarrow$ $0.04 \times 400(200 - x) = 0.05 \times 4{,}200(x - 10) + 0.06 \times 400(x - 10)$

$\Leftrightarrow$ $3{,}200 - 16x = 210x - 2{,}100 + 24x - 240$

$\Leftrightarrow$ $5{,}540 = 250x$

$\therefore$ $x = 22.16^0C$

8. a. i. Differentiate between heat and temperature.

ii. What is the purpose of the constriction in clinical thermometer?

iii. When a beaker of water is heated at the top, convection does not occur. Explain this observation.

iv. State two similarities between boiling and evaporation.

b. A 125W heater and a thermometer were immersed in 0.6kg of oil in a vessel of negligible heat capacity. The following results in the table were obtained.

Time (min)	2	4	6	8	10
Temperature (Kelvin)	294	302	313	324	334

i. Plot a graph of temperature (y-axis) against time on the grid provided.

Use your graph to determine:

ii. The average rise in temperature per minute

iii. The temperature at which heat started.

iv. The specific heat capacity of the oil.

c. What precaution should be taken in the experiment above to ensure that the results obtained are as accurate as possible?

Solution:

a. i.
- *Heat is a form of energy flowing from one point to another due to temperature difference while temperature is the degree of hotness or coldness.*
- *Heat is measured in Joules while temperature is measured in Kelvin.*

ii.
- *Controls the movement of thermometric fluid.*
- *Enables temperature to red at any time.*

iii.
- *Convection goes up- water is a poor conductor.*
- *Hot water has lower density than cold water.*

iv.
- *Both occur due to change in temperature*
- *In both there is vapour.*

b. *i.*

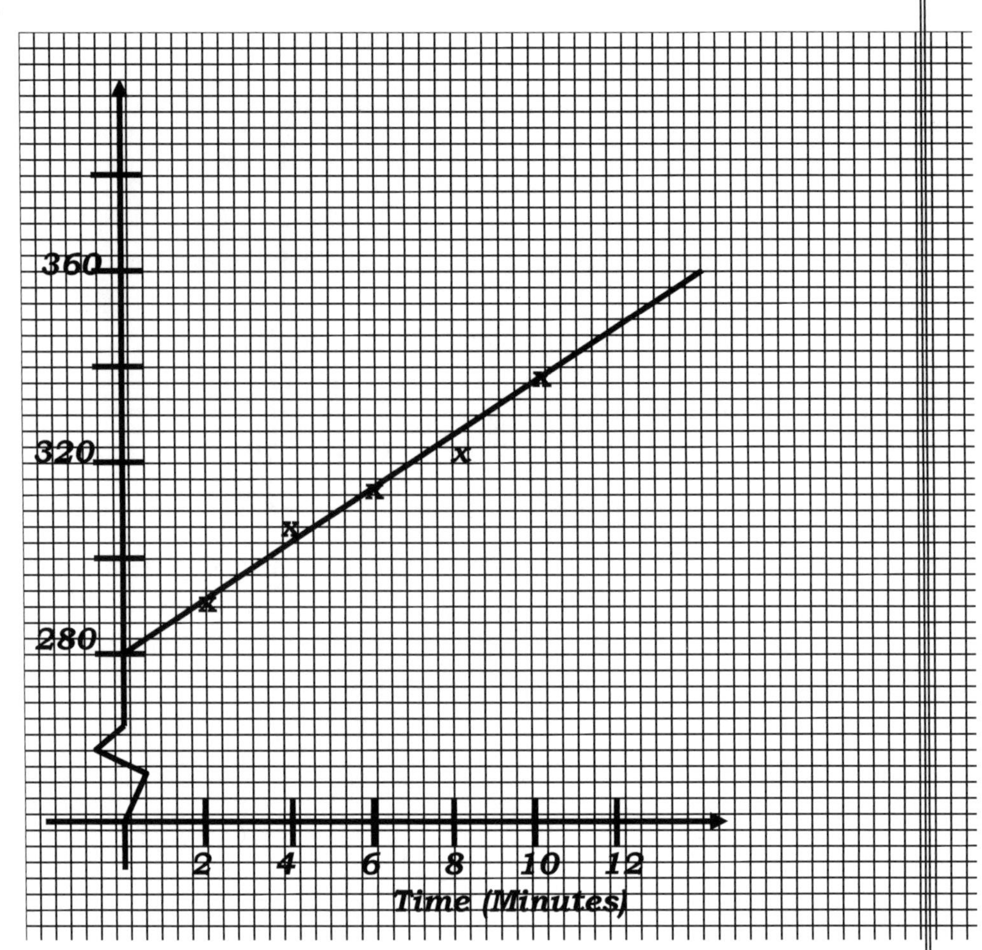

ii. Average rise in temp. = Slope = $\frac{334 - 294}{10 - 2} = \frac{40K}{8min} = 5K/min$

iii. 280K

iv. Heat $Q = Pt = mc\theta$

$\therefore \quad C = \frac{Pt}{\theta}$ and $\frac{\theta}{t} = 5K/min = 5K/60s$ (slope)

$\Rightarrow \quad C = \frac{125}{0.6} \times \frac{60}{5} = 2{,}500Jkg^{-1}K^{-1}$

c. - Minimise heat losses
- Stirring to obtain uniform temperature.

9. a. i. The metal rod shown below is heated from end A.

A [rod] B

Sketch a graph of temperature against distance from end A

ii. A metal length 2m and coefficient of linear expansivity 1.0 x 10^5K^{-1} is heated from 20^0C to 70^0C. Calculate the increase in length of the metal.

b. i. Explain heat conduction in solids in terms of kinetic theory.

ii. A burn by steam is more severe than one by boiling water. Explain.

c. i. In hot deserts, people prefer their cars roof tops coated white.
Explain.

ii. State 4 differences between boiling and evaporation.

Solution:

a. i.

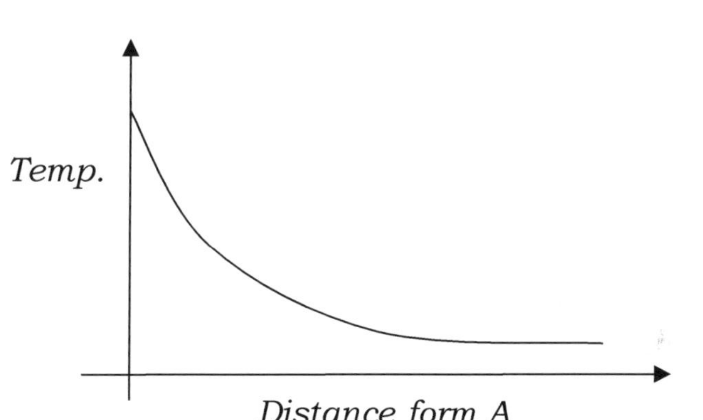

ii. Increase in length δL is given by

$\delta L = kL\delta\theta$

$= 1 \times 10^{-5} \times 2 \times (70 - 20)$

$= 1.0 \times 10^{-3} m$

b. i. Molecules vibrate strongly on heating solid. They knock against other molecules making them vibrate. Hot molecules therefore pass enough energy to cooler molecules.

ii. Steam has latent heat.

c. i. White roofs are poor heat absorbers.

ii. Differences:

- *Boiling takes place at specific temperature while evaporation takes place at all temperatures.*
- *Boiling takes place through out the liquid while evaporation takes place on the surface of liquid.*

- *In boiling heat energy is supplied while in evaporation heat energy requires not to be supplied.*
- *Evaporation causes cooling whereas in boiling temperatures remain constant.*

10. a. i. State two factors that affect the rate of evaporation.

ii. Why should the surface of a kettle be kept well polished?

b. To find the value of the specific heat capacity of a liquid, an electric heater is used to increase the temperature of the liquid in an aluminium can as shown in the diagram.

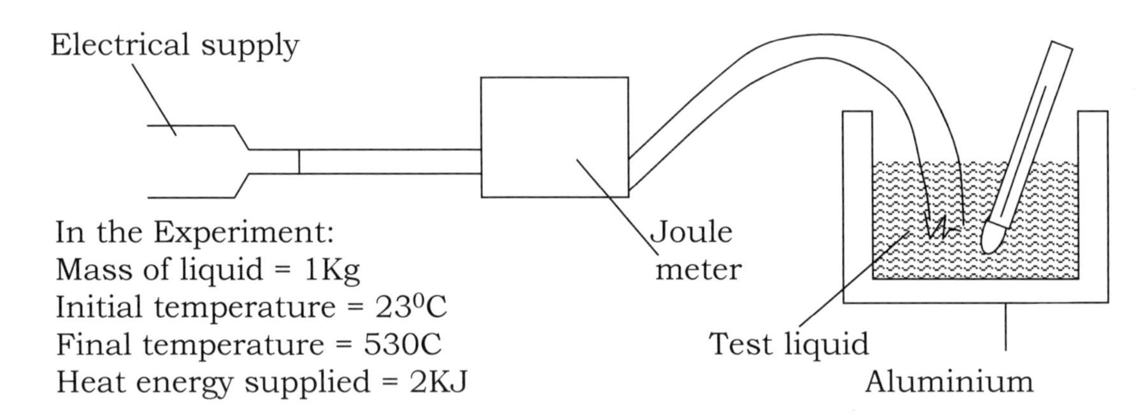

In the Experiment:
Mass of liquid = 1Kg
Initial temperature = 23^0C
Final temperature = 530C
Heat energy supplied = 2KJ

can

i. Determine the specific heat capacity of the liquid from these results.

ii. Suggest two sources of error in the experiment.

Solution:

a. i.
- *Increase in temperature increases rate of evaporation.*
- *Increase in surface area increases rate of evaporation.*
- *Increase in drought increases rate of evaporation*
- *Decrease in pressure above the surface of a liquid increases the rate of evaporation.*

ii. *To reduce heat loss by radiation.*

b. i. *Let the specific heat capacity be cJ/kg/^{0}C*

Heat gained = 1Kg x 300C x c = 24,000

$\therefore$ $c = \frac{24,000}{30} = 800J/kg/^0C$ *or* $800J/Kg/K$

ii. - *Aluminium can is not lagged hence much heat loss to the surrounding by radiation..*

- *The aluminium can is not covered to reduce evaporation.*

11. State two features in thermos flask, which reduce heat loss by radiation.

Solution:

1. Vacuum *2. Cork*

12. An electric immersion heater is dipped into a vessel containing 0.6kg of water. When switched on for two minutes, the temperature rises by 30^{0}C. Find the power of the heater. (Specific heat capacity, c, of water = 4,200$Jkg^{-1}K^{-1}$)

Solution:

$VIt = mc\theta \Rightarrow Pt = mc\theta$

$\Leftrightarrow 120P = 0.6 \times 4,200 \times 30$

$\Leftrightarrow P = \frac{0.6 \times 4,200 \times 30}{120}$

$= 630$ *watts*

13. The diagram shows the main compartment of a domestic refrigerator.

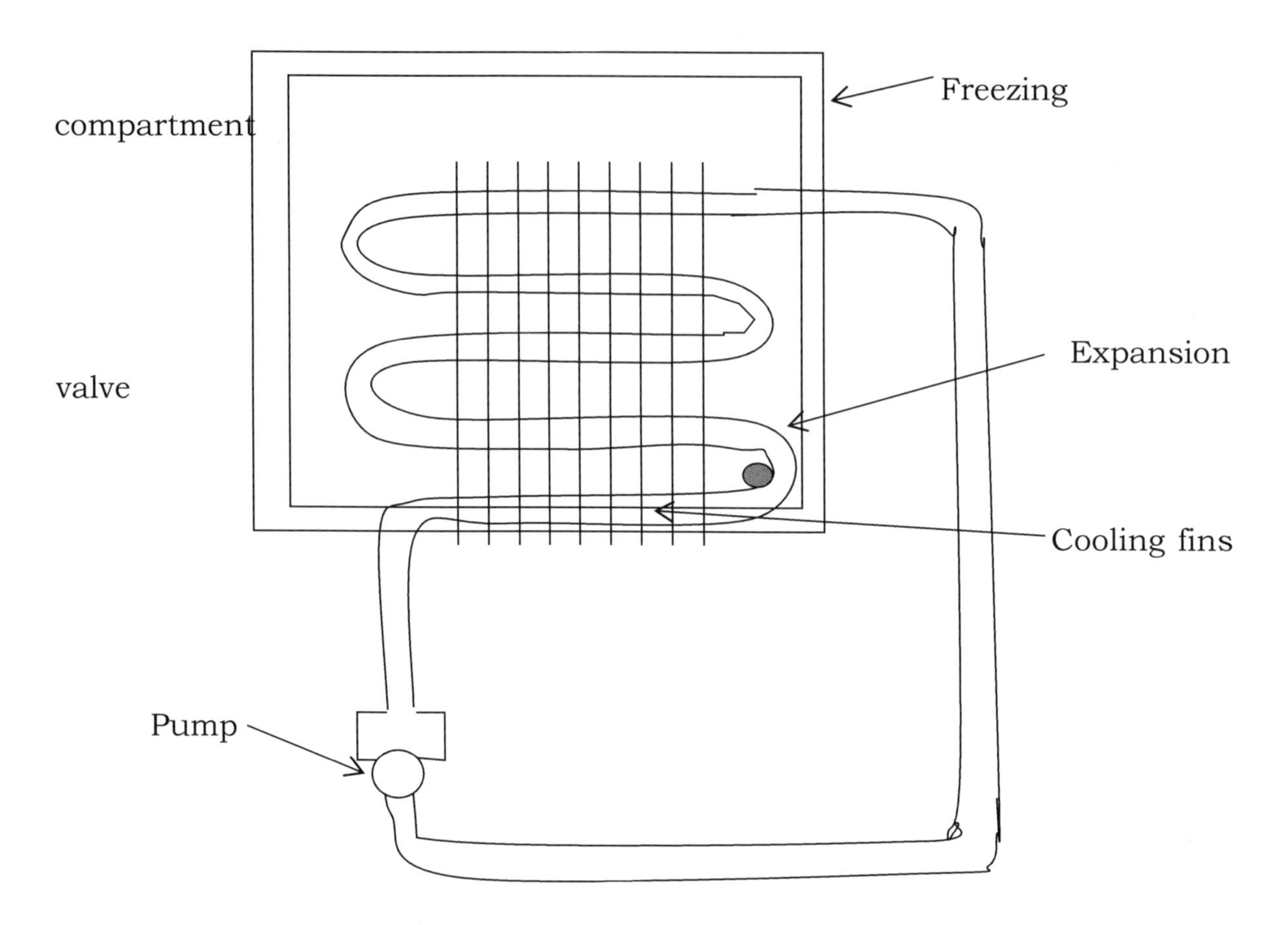

a. Why is the freezing compartment located near the top of the cabinet?

b. What is the purpose of the pump and the expansion valve?

c. Explain why the external metal fins become hot when the refrigerator is in operation.

d. When a tray of water at room temperature (20^0C) is placed in the freezing compartment, its temperature drops to freezing point (0^0C) in about 20 minutes. It takes a further 1½ hours before complete solid ice is formed in the tray. Explain why if is so.

e. It is required to convert 0.5kg of water at 20^0C into ice using a refrigerator, which can extract heat at an average rate of 20J/s. Determine whether this is possible within a period of 2 hours. (Specific heat capacity of water = 4,200J/kg; Specific latent eat of fusion of ice = 3.36×10^5J/kg/K)

Solution:

a. *The cooling tubes of the freezing compartment are near the top of the cabinet because by convection the warm air within the cabinet will be in contact with the tubes. As the air is cooled, the warmer air beneath it replaces it and the cabinet gets progressively cooler until the least possible temperature is attained.*

b. *The pump circulates the refrigerant. The expansion valve reduces the pressure on the freezer side of the system so that refrigerant evaporates becoming a vapour. On the higher-pressure side, the pump compresses the vapour so that it liquefies.*

c. *The pump compresses the vapour so that it liquefies in the tubes attached to the fins. As it liquefies, it gives out its latent heat. The metal tube and fins are good conductors of heat and are warmed up by the heat from the condensing refrigerant. I.e. the heat is being removed by the cooling fins.*

d. *Specific latent heat of fusion is very large and a lot of heat has to be removed before all water present gets frozen to solid ice. The heat needed to cool water from 20^0C to 0^0C is comparatively very little and hence does not take long.*

e. *Heat required to reduce temperature from 200C to 00c is given by*

$Q = mc\theta$

$= 0.5kg \times 4,200J/kg \times 20K$

$= 42,000J$

Heat lost by water at 00C, changing from liquid to solid is given by

$Q = mL$

$= 0.5kg \times 336,000J/kg$

$= 168,000J$

$\therefore$ *Total heat withdrawn = 42,000 + 168,000 = 210,000J*

Heat extracted in 2 hours = 20J/s x 2 x 3,600s

= 144,000J

Since less heat is extracted in 2 hours, all the water will not be converted into ice.

14. Give a reason why a concrete beam reinforced with steel does not crack when subjected to changes in temperature.

Solution:

Concrete and steel have the same linear expansivity.

Chapter 16: Radioactivity

1. a. Four nuclides are represented by the following symbols

$$^{54}_{27}A \qquad ^{59}_{29}B \qquad ^{58}_{30}C \qquad ^{58}_{29}D$$

i. Which nuclides are isotopes of the same element?

ii. Name the nuclides one of which could be produced from the other by emissions of a β-particle. Write the equation of the reaction.

iii. State two uses of radioactive emissions in everyday life.

b. In an experiment to determine the half-life of a radioactive element R,

the following data was obtained.

Activity (Counts/sec)	42	33	24	19	14	10	8
Time (sec)	0	1	2	3	4	5	6

iv. Draw a decay curve and use it to estimate the half-life of the element R.

v. Determine the decay constant of the element R

Solution:

$$^{59}B \qquad ^{58}_{29}D$$

29

a. i. and

N.B. By definition isotopes of an element are atoms, which have the same atomic number but different mass numbers.

ii. D and C

$$^{58}_{29}D \xrightarrow{\beta} {}^{58}_{30}C + {}^{0}_{-1}e$$

iii. <u>Uses:</u>
In medicine. Industries, Agriculture, Mutation (biology), Sterilisation of equipments etc.

b.

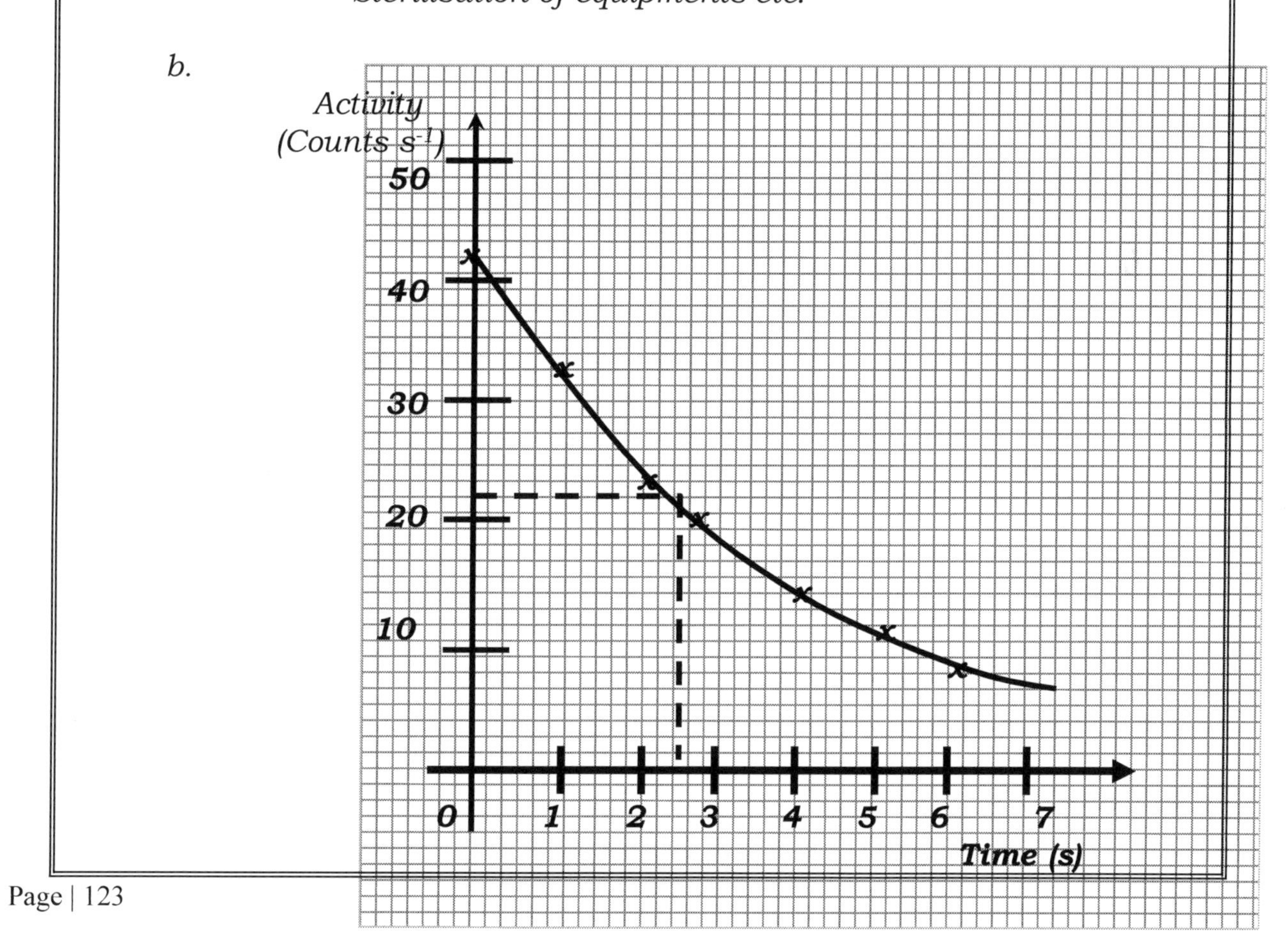

Half-life: *Original counts of 42*

$\therefore$ *Half value of count 21 and 2.4 seconds corresponds*

(From the graph)

ii. *Decay constant,* $\lambda = \frac{0.693}{\frac{1}{2} \times 2.4} = 0.577$

2. a. Define radioactivity

b. The count rate from a radioactive element is measured daily. The results are as shown in the table.

Time (days)	0	1	2	3	4	5	6
Count Rate (per minute)	12,000	7,560	5,000	3,300	2,150	1,400	900

i. Plot a graph of count rate against time.

ii. From the graph, determine the half-life of the element.

Solution:

a. *Radioactivity is the spontaneous disintegration of the nucleous of a radioactive element.*

b.

3. Find the value of **Y** in the reaction represented by the following equation.

$$^{210}_{83}\text{Po} \xrightarrow{\beta} {}^{210}_{\mathbf{Y}}\text{Po}$$

Solution:

Y = 84

4. A radioactive substance as a half-life of 20 minutes. If the initial mass of the substance is 2 grams, how much of it has decayed in one hour.

Solution

Initial mass m = 2g ⇒ after 20 min ⇔ ½ x 2g = 1g

⇒ after 20 min ⇔ ½ x 1g = 0.5g

⇒ after 20 min ⇔ ½ x 0.5 = 0.25g

∴ mass decayed after 1hr = 1.75g

5. A radioactive isotope of half-life of 4 years was bought in 1990. If its mass in 2006 is 1g, how much of the isotope was purchased originally?

Solution:

$$\text{No. of half-lives} = \frac{2006 - 1990}{4} = \frac{16}{4}$$

$$= 4 \text{ years}$$

$$\text{Mass remaining} = (^1/_2)^4 = ^1/_{16}$$

$\Rightarrow$ $1/16 \rightarrow 1g$

$\therefore$ $\text{Mass purchased} = \dfrac{1}{^1/_{16}} = 16g$

OR

$2006 \rightarrow 2002 \rightarrow 1998 \rightarrow 1994 \rightarrow 1990$

1g 2g 4g 8g 16g

$\therefore$ *Mass purchased = 16g*

6. Explain why radiocarbon dating cannot be used to estimate the age of a rock.

Solution:

- *Radiocarbon dating work by measuring the decay of carbon 14, which is found in living things. Rocks have never lived so they do not contain any carbon.*

7. What is half-life of a radioactive substance?

Solution:

- *It is the period/time taken by a radioactive substance to decay by half its original value.*

Chapter 17: Waves and Sound

1. a. Sketch a transverse wave and on it show:
   i. The wave length
   ii. The amplitude of the wave.

   b. A student timed the interval of his heartbeat and found that it took 0.5 seconds between any two successive beats. Find
   i. The frequency of his heart beats.

   ii. If the speed of the heartbeat was found to be $340ms^{-1}$, calculate the wavelength of the wave produced by his heartbeat.

   c. Two students are provided with a tape measure, a pair of wooden blocks, some chalk powder and stop watch. Briefly explain how they can use the apparatus to determine the speed of sound in air.

   d. Define the term echo.

Solution:

a.

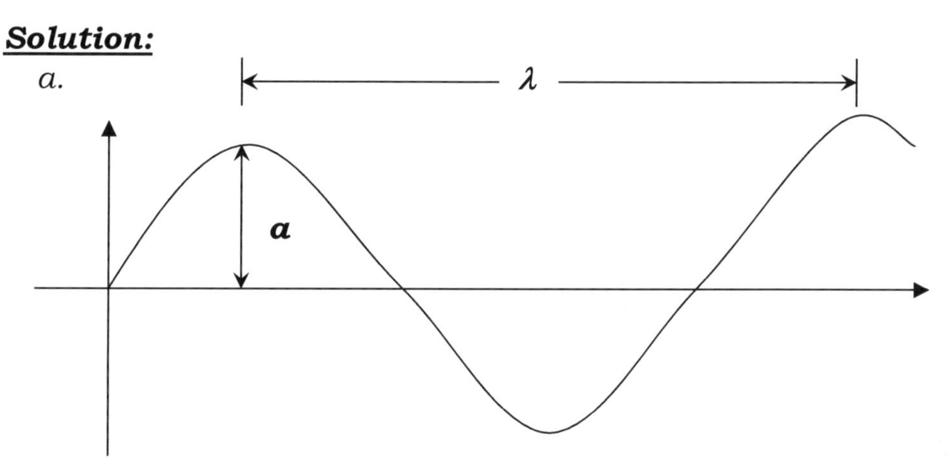

b. i. $f = 1/T = 1/0.5 = 2Hz$

ii. $\lambda = v/f = 340/2 = 170m$

c. *Let the student with the stopwatch stand at a distance from the one with the clapper. Measure the distance between the two students. Let the student with the clapper clap and the other measures the time intervals from when he sees the dust to the time he hears the clap.*

Calculate the speed, s, of the sound in air as:

$$S = \frac{distance}{Time\ taken}$$

d. *Echo is reflected sound.*

2. The figure represents crests of straight waves produced on a ripple tank.

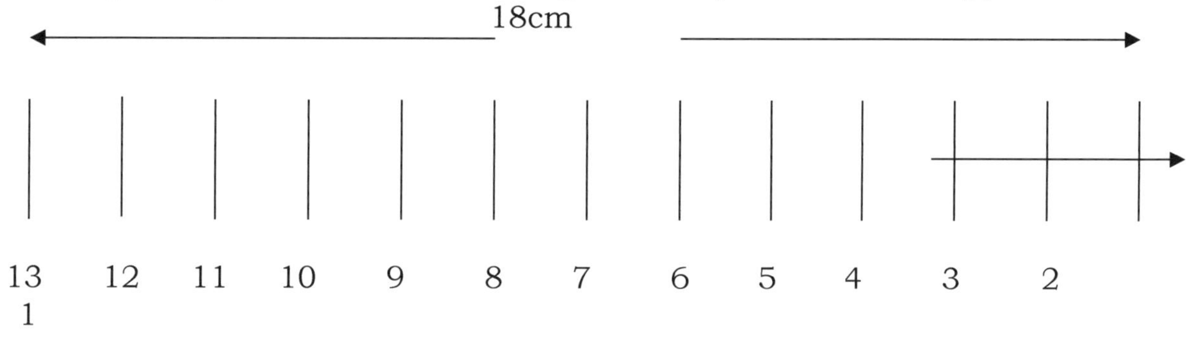

a. Determine the wavelength of the waves.

b. Given that crest number 1 occupied the position now occupied by crest number 13, 14.40 seconds ago, determine the frequency of the waves.

Solution:

a. *Wavelength* = $\frac{\text{length of all waves}}{\text{No. of waves}}$

$= \frac{18\ cm}{12} = 1.5cm$ *Or* $0.15m$

b. *Wave velocity* = $\frac{\text{Distance}}{\text{Time}}$

$= 1.8cm/14.40s = 1.25cm/s$

$= 0.0125m/s$

3. i. Distinguish between stationary waves and progressive waves giving one example for each.

ii. State two quantities that change as wave travels from one medium to another.

Solution:

i. *Progressive waves:- associated with transmission of energy from one point to another but stationary waves are not accompanied by transfer of energy.*

ii. *Velocity and wavelength*

4. a. The figure, not drawn to scale, represents an experiment to show Young's fringes using a monochromatic red light.

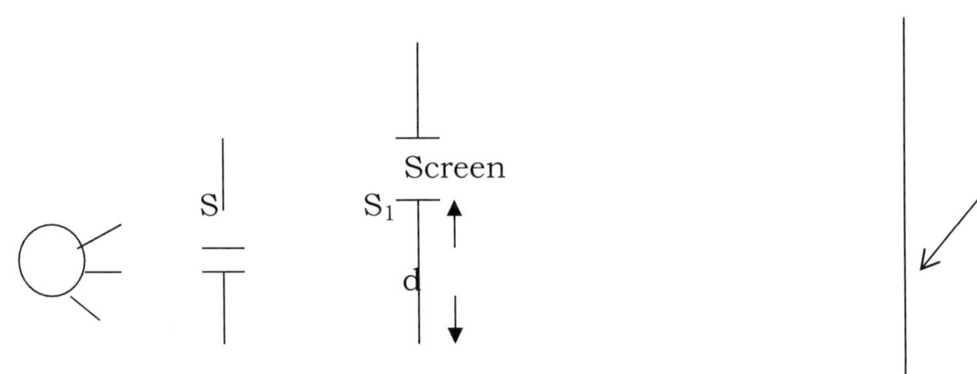

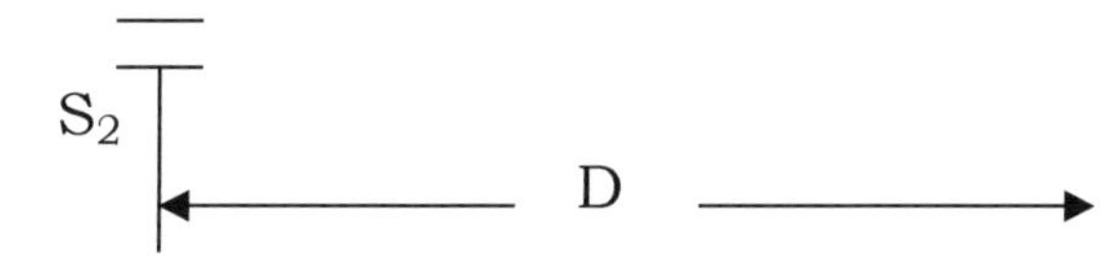

i. What is the purpose of the single slit S?

ii. What happens to the fringes patterns if the distance d is increased?

iii. What would happen if the slit S_2 is blocked leaving S and S_1 unchanged?

iv. What could be the observation made if the red light is replaced with white light?

b. A guitar string is sounded together with a tuning fork of frequency 396Hz produced 6 beats per second. The string was then tightened and produced 3 beats per second.

i. What is the possible initial frequency of the guitar string?

ii. What is the new frequency of the string?

c. State the conditions necessary for two progressive waves travelling in opposite direction to produce a stationary wave.

Solution:

a. i. *To act as a coherent source of the light.*

ii. *The fringes on the screen move closer.*

iii. *No fringes would be found.*

iv. *Bright spectrum alternating with dark bands would be formed.*

b. i. $f_0 = 396 \pm 6 = 402Hz$ *or* $390\ Hz$.

ii. *New frequency* $f_f = 396 - 3 = 393Hz$.

c. *They must have equal speed, equal frequency and nearly equal amplitude.*

5. a. The following shows a displacement time graph.

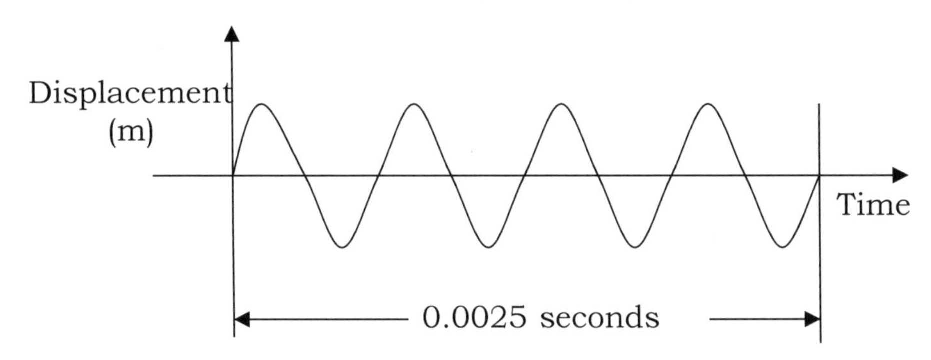

What is the frequency of the wave shown?

b. What is the main difference between transverse and longitudinal waves?

c. A student standing between two vertical cliffs and 640m from the nearest cliff shouted. She heard the first echo after 4 seconds and the second echo 6 seconds later. Using this information, calculate

i. The velocity of sound in air.

ii. The distance between the cliffs.

Solution:

a. *4 waves ⇒ 0.0025s* $\quad \therefore$ *1 wave ⇒* $\frac{0.0025}{4}$

$f = \frac{1}{T} = \frac{4}{0.0025} = 1{,}600Hz$

b. *Transverse wave – the vibrations of particles are at right angles to the direction of wave-motion.*

Longitudinal wave – the vibrations of particles are parallel to the direction of the wave-motion.

c. i. *Velocity* $= \frac{displacement}{time}$

$= \frac{640 \times 2}{} = 320m/s$

4

ii. *Distance = speed x time*

= 320 x 10/2 = 1,600m

6. A boy is standing in front of a cliff. He claps his hands and hears an echo 2 seconds later. Find the distance between the boy and the cliff.

Solution

Distance = speed x time

= 320 x 2 x ½

= 320m

7. The figure is a display of a waveform, 0.002 seconds after it is released form the source.

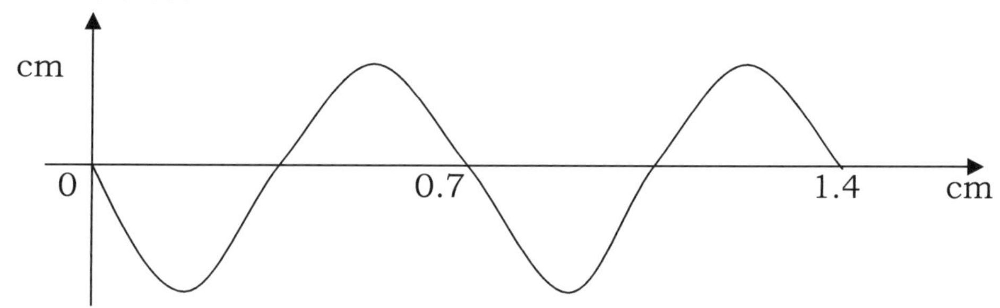

Determine:

a. The wavelength

b. The frequency of the wave.

Solution:

a. *Wavelength = 0.7cm*

b. $f = \frac{1}{T}$, where $T = \frac{0.002}{2} = 0.001sec$

$f = \frac{1}{0.001} = 1000Hz$

8. Distinguish between refraction and diffraction as used in waves.

Solution:

- *Refraction is the bending / changing of direction by a wave as it crosses two media of different densities.*

> *Diffraction is the spreading of wave round a barrier after passing through a narrow opening.*

9. State the effect of sound moving from a hot air region to cold air region on its:
   i. Wavelength
   ii. Frequency

Solution:
- *Wavelength increases.*
- *Frequency remains unchanged.*

10. A boy whistles while standing 150m from a high wall. A girl standing X meters behind the boy hears the echo 1.3 seconds later. Determine the value of X.

Solution:

$$t = \frac{d}{S} \Rightarrow \frac{(150 \times 2) + X}{330} = \frac{300 + X}{330} = 1.3$$

$$\Leftrightarrow \quad 300 + X = 429$$

$$\therefore \quad X = 129m$$

11. It is observed that as sound waves travel from a dense to a less dense gas, its velocity changes. Which wave property does this show?

Solution:
Refraction of waves as they move from one medium to another of different densities.

12. Water ripples are caused to travel across the sallow surface of a tank by means of a suitable vibrator. The distance between two successive crests is 3.0cm and the wave travels 25.2cm in 1.2 seconds. Calculate the frequency of the vibrator.

Solution:

$$f = \frac{v}{\lambda} \quad \text{but} \quad v = \frac{0.252}{1.2} = 0.21m/s$$

$$\text{and} \quad \lambda = 3.0cm = 0.03m$$

$$\therefore \quad f = \frac{0.21}{0.03} = 7Hz$$

13. Explain the term “Resonance”.

Solution:

Resonance occurs when a body vibrates at its natural frequency as a result of vibrations from another system vibrating with the same frequency.

14. Name one piece of evidence that shows electromagnetic waves are transverse.

Solution:

They can be polarized, which is only possible with transverse waves.

15. What happens to the speed of sound in air if the temperature rises? Give reasons.

Solution:

- *The speed of sound increases.*
- *At high temperatures, air becomes light/less denser.*

16. The figure shows some water waves travelling at a frequency of 50Hz. Find the velocity of the water waves.

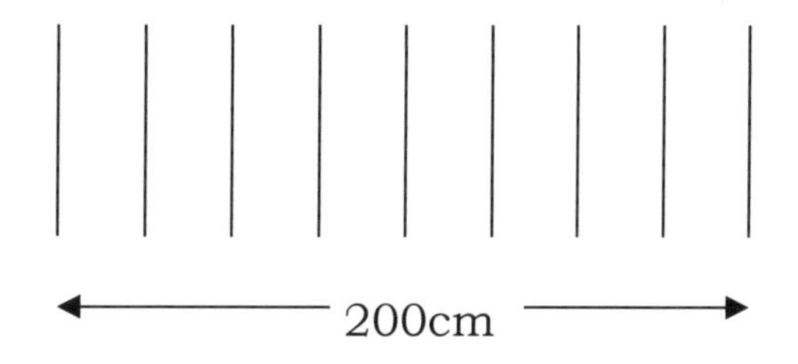

200cm

Solution:

$v = \lambda f$

but $\lambda = \frac{200}{8} = 25cm = 0.25m$

$\therefore \quad v = 0.25 \times 50 = 12.5m/s$

17. The figure shows how water spread out after passing through a gap.

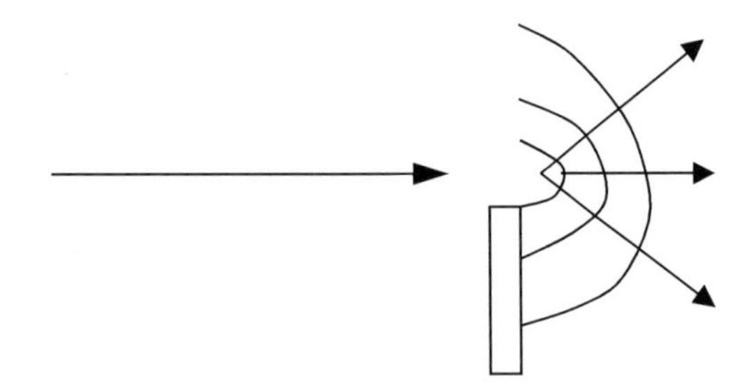

What two factors determine the degree of spreading of the waves?

Solution:

- *Wavelength*
- *Size of the aperture*

Manufactured by Amazon.ca
Bolton, ON

26886152R00074